AN INTRODUCTION TO MICROELECTRONIC TECHNOLOGY

AN INTRODUCTION TO MICROELECTRONIC TECHNOLOGY

D. V. Morgan
University of Leeds

K. Board
University College, Swansea

Adapted by

Richard H. Cockrum
California State Polytechnic University

John Wiley & Sons
New York • Chichester • Brisbane • Toronto • Singapore

Copyright © 1985, by John Wiley & Sons, Inc.

All rights reserved. Published simultaneously in Canada.

Reproduction or translation of any part of
this work beyond that permitted by Sections
107 and 108 of the 1976 United States Copyright
Act without the permission of the copyright
owner is unlawful. Requests for permission
or further information should be addressed to
the Permissions Department, John Wiley & Sons.

Library of Congress Cataloging in Publication Data:

Morgan, D. V.
   An introduction to microelectronic technology.

  Bibliography: p. 195
  Includes indexes.
   1. Microelectronics. I. Board, K. II. Cockrum,
Richard H. III. Title.
TK7874.M53357 1985   621.381'7   84-26951
ISBN 0-471-81073-8

Printed in the United States of America

10 9 8 7 6 5 4 3 2 1

I dedicate this adaptation to my daughter
Tawny Lynn

PREFACE

This textbook is written primarily as a "first exposure" to the topic of microelectronics. The discussion is organized so that it is suitable for a wide range of students, both in two- and four-year programs. Students who wish to be introduced to semiconductor microelectronics using a minimum of mathematics can simply study the material presented in Chapters 1 to 9, answering the problem sets at the end of each chapter. A set of self-evaluation questions is presented at the beginning of each chapter, highlighting the concepts that are important to understand in that chapter. Chapter 1 discusses the development of microelectronics. Chapter 2 describes the basic materials used to make semiconductors. Chapter 3 explains how the basic materials are modified to meet the specific electrical requirements. Chapter 4 deals with oxidation, Chapter 5 with lithography, and Chapter 6 with metallization, interconnections, and bonding. Chapters 7 and 8 describe the actual design and fabrication of an integrated circuit. Chapter 9 speculates on the future trends in microelectronics. The more advanced student should study Appendixes 1, 4, and 5 and answer the problems in Appendix 7. These appendixes are intended for students who have completed a beginning course in calculus. The organization of the book makes it easy to select the level of mathematics desired without having to sort out which sections of each chapter might be suitable.

I wish to acknowledge D. V. Morgan of the University of Leeds and K. Board of the University College, Swansea as the authors of the manuscript from which this adaptation was made. I thank C. J. Skinner for typing this manuscript.

Pomona, California 1984   *Richard H. Cockrum*

LIST OF SYMBOLS

B	Magnetic field strength
C	Capacitance
C_f	Correction factor for 4-point probe
c	Velocity of light
D	Diffusion constant of impurity
D_0	Diffusion coefficient of oxidant in SiO_2, Diffusion coefficient at $T \to \infty$
d	Junction depth
E_a	Diffusion activation energy
E_g	Band gap energy
E_s	Formation energy of a vacancy
E_1	Activation energy of oxidant diffusing in SiO_2
E_2	Activation energy of reaction to form SiO_2
F	Electric field strength
F_f	Oxidant flux
F_i	Oxidant flux at SiO_2 interface
F_0	Oxidant flux through SiO_2
f	Resultant force on charge carrier
I, I_x, I_y, I_z	Current and its cartesian coordinates
K_1, K_2	Constants relating to the thermal oxidation process
k	Boltzmann's constant
k_i	Reaction rate constant of thermal oxidation process
m^*	Effective mass of charge carrier (m_e^* electrons, m_h^* holes)
N	Oxidant concentration, Impurity concentration
N_A	Acceptor concentration
N_D	Donor concentration
N_i	Oxidant concentration at interface
N_m	SiO_2 molecules per unit volume
N_0	Surface atomic concentration
N_s	Oxidant concentration at oxide surface
N_T	Total impurity concentration

List of Symbols

n	Free electron concentration
N_1, N_2	Volume concentration of atoms
n_1, n_2	Atomic density in atomic planes 1 and 2 respectively
n_i	Intrinsic carrier concentration
p	Free hole concentration
Q	Area density of atoms diffusing into solid
q	Electronic charge
R	Resistance
R_H	Hall coefficient
R_p	Mean projected range of implanted ion
$R_\perp$	Mean transverse range of implanted ion
S	Device area
T	Absolute temperature
T_f	Freeze-out temperature
t	Time
u	Mean drift velocity
V	Potential
V_T	Threshold voltage (enhancement MOST)
v, v_x	Particle velocity
x_0	Oxide thickness
x_i	Initial oxide thickness
ε_s	Permittivity of silicon
ϕ	Net flux of diffusing atoms
λ	Mean free path between collisions
μ	Mobility
μ_H	Hall mobility
μ_M	GMR mobility
ν	Atomic vibrational frequency
ν_j	Atomic jump frequency
θ_p	Range straggling
ρ	Resistivity
$\rho_\square$	Sheet resistivity
σ	Conductivity
τ	Mean time between collisions
τ_o	Initial oxidation time
χ	Electron affinity

CONSTANTS

q	electronic charge	1.602×10^{-19}	coulombs
ϵ_0	permittivity of free space	8.85×10^{-14}	farad/cm
k	Boltzmann's constant	8.62×10^{-5}	eV/K
h	Plank's constant	6.63×10^{-34}	j-sec
m_0	electron rest mass	9.11×10^{-31}	kg
μ	micrometer	10^{-4}	cm
eV	electron volt	1.602×10^{-19}	joule

CONTENTS

List of Symbols   ix
Constants   xi

1 The Development of Semiconductor Technology   1
Instructional Objectives   1
Self-Evaluation Questions   1
1.1   Introduction   1
1.2   Advantages of Integration   2
1.3   Levels of Integration   3
1.4   The Problem of Large Volumes   6
1.5   Batch Processing of a Prototype Integrated Circuit   8
Problems   11

2 Materials for Semiconductor Devices   13
Instructional Objectives   13
Self-Evaluation Questions   13
2.1   Introduction   14
2.2   Crystal Purification and Growth   15
2.3   Epitaxial Growth   22
2.4   Chemical Etching   32
Problems   35

3 Impurity Doping in Semiconductors   37
Instructional Objectives   37
Self-Evaluation Questions   37
3.1   Basic Concepts   38
3.2   Type Conversion   49
3.3   Atomic Diffusion   50
3.4   Ion Implantation Doping   56
Problems   67

4 Insulating Films on Semiconductors 69
Instructional Objectives 69
Self-Evaluation Questions 69
4.1 Introduction 70
4.2 Types of Surface Oxides 71
4.3 Practical Oxidation Systems 73
4.4 Mechanisms and Design Rules for Thermal SiO_2 Film 73
4.5 Additional Effects in the Oxidation Process 78
4.6 Assessment of Film Quality 80
4.7 Choice of Film Type 81
4.8 Nitride Films 81
Problems 84

5 Photolithography 85
Instructional Objectives 85
Self-Evaluation Questions 85
5.1 Introduction 85
5.2 Photoresist Types 88
5.3 Film Thickness 90
5.4 Masks and Mask Making 93
Problems 95

6 Metallization, Interconnections, and Packaging 97
Instructional Objectives 97
Self-Evaluation Questions 97
6.1 Introduction 98
6.2 Thin Film Deposition 98
6.3 Plating 103
6.4 Metallization System 104
6.5 Surface Protection and Wafer Thinning 106
6.6 Dicing, Mounting, and Bonding 107
Problems 115

7 Fabrication of Devices and Circuit Components 117
Instructional Objectives 117
Self-Evaluation Questions 117

7.1 Introduction **118**
7.2 Simple *pn* Junctions **118**
7.3 Bipolar Transistors **124**
7.4 Junction FET (JFET) **126**
7.5 The Metal Semiconductor FET (MESFET) **127**
7.6 The Metal Oxide Semiconductor Devices (MOS) **129**
7.7 Charge Coupled Devices **131**
7.8 Passive Circuit Elements: Resistors and Capacitors **133**
7.9 Gallium Arsenide ICs **136**
7.10 Special Device Structures **138**
Problems **140**

8 MOS and Bipolar Technologies and Their Applications 141

Instructional Objectives **141**
Self-Evaluation Questions **141**
8.1 Introduction **142**
8.2 Technology Families **142**
8.3 Silicon-Gate *n*MOS Process **148**
8.4 The Bipolar Process **150**
8.5 Integration of a Commercial Circuit **153**
Problems **158**

9 Future Developments in Semiconductor Technology 159

Instructional Objectives **159**
Self-Evaluation Questions **159**
9.1 Technology: The State of the Art **159**
9.2 Future Technology Developments **162**
9.3 Conclusions **165**
Problems **167**

Appendix 1. Material and Device Evaluation Technologies **169**
Appendix 2. Common Etches for Silicon **181**
Appendix 3. Common Etches for Gallium Arsenide **183**

Appendix 4. The Mathematics of Diffusion and Ion Implantation 185
Appendix 5. Thermal Oxidation of Silicon 193
Appendix 6. References and Bibliography 195
Appendix 7. Supplemental Exercises 197
Appendix 8. Properties of Si, GaAs, and SiO_2 203

Index 205

CHAPTER 1
The Development of Semiconductor Technology

Instructional Objectives

This first chapter acquaints the student with the basic reasons for producing integrated circuits (ICs). It assumes the student already knows, in general, the definition of an IC. After reading this chapter, you should be able to:

a. Explain the cost advantage of integrated circuits over discrete circuits.
b. Explain some of the problems associated with ICs, such as the yield.
c. Explain a simple batch processing technique for making ICs.

Self-Evaluation Questions

Watch for the answers to these questions as you read the chapter. They will help point out the important ideas presented.

a. What is the typical diameter of silicon wafers used to make ICs?
b. What are the cost advantages of ICs over discrete circuits?
c. What are the factors that limit the maximum complexity of ICs?

1.1 Introduction

With the invention of a solid state amplifying device in 1947 by Shockley, Bardeen, and Brattain, the possibility of "integrating" not one, but many transistors within the same crystal has existed. Previously, the vacuum tube provided the only means of amplifying an electrical signal, and integration in this context was never feasible.

The economic advantages and improvements in performance that occur in integrating many devices within the same package or chip are, as we discuss later in this chapter, very great. There has thus grown up over the last 20 years a technology that has enabled increasing numbers of devices to be placed on a single chip.

Sophisticated new techniques have been developed and new ones are emerging in an effort to continue this trend toward higher integration levels. This has reached the point where a new discipline, that of "semiconductor microtechnology" has emerged and which constitutes the main purpose of this book.

The development of this technology has been swift but expensive. Since the first "integrated circuit" in 1959, the number of components per chip has almost doubled every year so that it is now well in excess of 100,000. The pressures to bring about this revolution in electronics have to be powerful indeed and most of this chapter is devoted to a discussion of the reasons for it. In the final part of the chapter, a simple prototype integrated circuit is discussed in detail. Although the circuit itself is simple, its fabrication involves most of the processing steps found in more complicated integrated circuits and it will, therefore, serve to illustrate the individual process and how they are linked together to produce an overall sequence.

1.2 Advantages of Integration

An interesting analogy may be drawn between the development of the integrated circuit and the automobile. Society has been changed profoundly by both over a relatively short period and the reasons in both instances were primarily economic.

It was not the invention of the internal combustion engine in 1884 that initiated the large-scale use of the automobile in society. For many years it was only available to a small number of very wealthy people. Rather, it was the invention in 1903 of mass production techniques by Henry Ford that gave rise to its widespread use.

Over the last few years, the impact of electronics on all aspects of our society has been rapid and dramatic, although the basic electronic functions and operations were known for many years before this. The reason is that a technique known as "batch processing" has made electronic functions and operations far cheaper than before. The technique has similarities with mass production and has had similar consequences. The dramatic reduction in the cost of electronic functions has, in parallel with the automobile, made it available to people and applications that would have been ruled out for economic reasons.

1.2.1 BATCH PROCESSING

It is readily possible to make even complex circuits on areas of silicon less than 5 mm square. Typically, silicon wafers of 75 to 100 mm (3 to 4 in.) in diameter are used so that each contains about 180 to 500 circuits. Each wafer is passed through the various processing stages so that each circuit is treated simultaneously or as a batch. Furthermore, it is usual for the manufacturer to process several hundred wafers together (i.e., up to 100,000 circuits). Consider, for example, one of these processing stages known as diffusion (described in detail in Chapter 3).

If it costs $10,000 to run the diffusion equipment, pay the salaries for that run, and 200 75 mm (3 in.) wafers each containing 180 circuits are processed together, the cost per circuit is only 10,000 ÷ (200 × 180) = 0.28, or about $0.28. (Actually, one should consider here only the number of "working circuits, since some will malfunction; this problem of "yield" is discussed later in the chapter.) Although the actual cost of mass production is high, the added cost per circuit is very low.

There are, of course, stages in the manufacture of the circuit that must be carried out on each circuit individually; these include packaging and testing. The costs are not divided by the number produced but have to be added directly to the unit cost. Thus, they are usually the most important component in the cost of present-day integrated circuits.

1.3 Levels of Integration

The more components manufacturers are able to place on a single semiconductor chip, the higher the level of integration they can achieve. At one extreme, they may use single transistors, each separately packaged, and separate resistors, capacitors, and other components. This is the fully "discrete" approach. At the other extreme, manufacturers may place the total electronic system on the chip and thus achieve the highest possible level of integration. There are a whole range of levels between these limits, and the decision on which is optimum is a complex economic one, depending on factors such as the availability of integrated circuits and projected market size.

TABLE 1.1 Cost Comparison of the Three Different Levels of Integration

Unit Cost	Discrete Approach	Partly Integrated Approach	Fully Integrated
Design costs	$80K/10^6 = $0.01	$80/10^6 = $0.01	$80K/10^6 = $0.01
Component costs	10,000 @ $1.0 = $10K	100 ICs (100 comp/chip) @ $6 = $600	1 IC @ $20 = $20
PC board	200 boards @ $10 = $2,000	2 PC boards @ $10 = $20	—
Assembly and packaging	$100	$40	$40
Total unit cost	$12,100	$660	$60

All prices indicated in this table are for example purposes only. Actual prices change almost monthly and vary from manufacturer to manufacturer.

1.3 Levels of Integration

To illustrate the powerful cost advantages of integration, we consider the example of an electronic system manufacturer who wishes to manufacture a system containing 10,000 components. His marketing manager advises him that 10^6 of these may be sold over a five-year period, after which the product is likely to be superseded by some competitor's product.

The total design costs for the system are $80,000. Every 50 components are connected together on a printed circuit board, and we assume that the latter costs $10 to produce. Table 1.1 shows three different approaches; one fully discrete, one fully integrated, and one intermediate case. The design costs are negligible, since they are disposed over 10^6 units (see also Figure 1.1). The individual "component" cost (here defined as a single packaged silicon chip) increases only slightly over the number of components on it; the major differences in the unit cost are due to the differences in the number of components and printed circuit boards required for each implementation.

Thus, it can be seen that there are very powerful cost incentives to

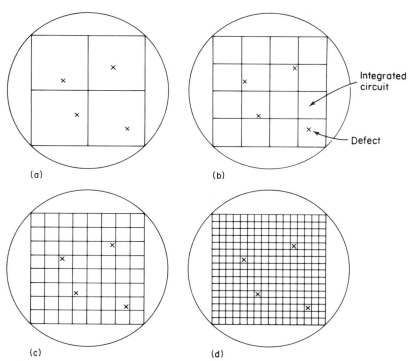

Fig. 1.1 Relationship between yield and chip size: (*a*) yield = 0; (*b*) yield = 75 percent; (*c*) yield = 94 percent; (*d*) yield = 98 percent.

move to higher levels of integration. In addition, there is a size and weight advantage, which is clear from Figure 1.1. Two-hundred printed circuit boards, with all their interconnections, would fill a large cabinet, whereas the two printed circuit board intermediate version would occupy only a small bench-top instrument casing. The final version could, depending on the peripherals required, fit in a hand-held case.

Although each of the three versions perform the same electronic function, there might be a performance advantage in the fully integrated form because of the elimination of long lead lengths. However, a much more important advantage is apparent in the latter implementation, reliability. The most common causes of failure tend to be the interconnections between the components and the rest of the circuit. Most of these interconnections have been placed on the chip in the fully integrated version and, therefore, eliminated, making this by far the most reliable, as well as the cheapest and smallest of the three versions considered.

In the rather oversimplified example just discussed, several factors that modify this simple picture should be mentioned.

1.4 The Problem of Large Volumes

In the previous example, we discussed the situation from the point of view of the electronic system manufacturer who purchased discrete or integrated circuits from an IC manufacturer. As far as the IC manufacturer is concerned, unless he or she can fill production lines close to full capacity, it is not worthwhile to produce them. If, for example, the manufacturer has a 75 mm (3 in.) wafer line with 10 production runs per week, with a circuit size of 5×5 mm (200 mils $\times$ 200 mils), he or she would produce 1.8 million annually (assuming only 10 percent of the output functioned correctly at the finish). The system manufacturer in our example only requires 2×10^5 annually for the product, so unless there were other customers, the IC manufacturer may not consider it worthwhile to produce the fully integrated version. The system manufacturer may then be forced to move to lower levels of integration.

There is a general rule that the more complex any system becomes, the more specialized it is and, hence, the smaller the number of applications. Thus, a move to higher complexity means that fewer are likely to be sold, which might mean in turn that they are not made because it is not economically feasible.

The microprocessor was developed specifically to circumvent this problem. It combines a high degree of complexity with "programability" so that it can be used to perform many different tasks. It has made it feasible for IC manufacturers to achieve high volume sales for very complex electronic systems and thus the systems and circuit designers have a powerful new tool at very low cost.

1.4.1 YIELD

The number of "working" circuits at the end of the process sequence is of particular concern to the integrated circuit manufacturer. If the process is new and innovative, then it is likely that in the early stages the yield will be low. The greater the number of stages in the process, the lower will be the yield, since each stage carries with it a finite probability that a malfunction will be induced in some of the circuits. For the case where the fraction still working after each stage f is the same for each stage, then the fraction working after N stages is $Y = f^N$. For example, for a 30-stage process with 10 percent lost at each stage ($f = 0.9$), the overall yield is only 4 percent. If the yield per stage f is 0.98 or only 2 percent lost at each stage, the overall yield increases to 54 percent. However, if the number of process steps, N, is increased to 50, then the yield drops to 36 percent with the same yield per stage.

1.4.2 MAXIMUM COMPLEXITY

Another factor that affects the decision on which integration level is appropriate is that there is, at a given moment, a maximum complexity that the IC manufacturer can achieve with adequate yields. At one end of the scale, he has a minimum feature size that sets a limit on the component density he can achieve. If he makes a circuit component smaller than what this minimum feature size will allow, he will suffer a dramatic decrease in yield because the misalignment between successive layers exceeds what is necessary to produce a working circuit.

At the other end of the scale, there is a maximum chip size above which the yield again produces rapidly. The starting semiconductor substrate or wafer has a fixed number of defects per unit area, as is illustrated in Figure 1.2. If we assume that each defect is effective in preventing the chip that incorporates it from working, then the maximum yield that the manufacturer can achieve is zero with four chips per wafer, 75 percent with 16, and 98 percent with 256. It would be inadvisable for him to operate any process at a chip size corresponding to 4 per wafer, since his yields are severely limited before he starts.

Both factors considered together mean that there is a maximum achievable complexity in terms of the number of components per chip. This can only be increased by

8   The Development of Semiconductor Technology

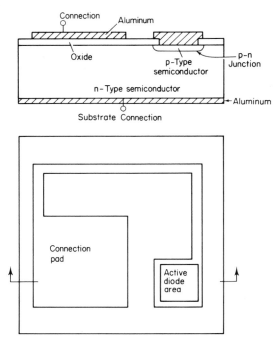

Fig. 1.2 A simple prototype integrated circuit to illustrate the various processing stages.

a. The supplier of the substrates producing lower defect densities in his starting material.
b. Obtaining new processing equipment that permits a smaller minimum feature size.

Both are outside the IC manufacturer's control and are improvements that only come slowly with time and at considerable expense.

Subsequent chapters in this book are concerned with specific stages in the fabrication of integrated circuits. To give an overall view of the complete technology, we devote the rest of this chapter to a discussion of a simple prototype IC which, although in circuit terms is unrealistic as a commercial IC, nevertheless contains all the individual process steps used in much more complex circuits.

1.5 Batch Processing of a Prototype Integrated Circuit

A cross section and plan view of the circuit, in this case is a simple *pn* junction diode, is shown in Figure 1.2. The connection pattern consisting

1.5 Batch Processing of a Prototype Integrated Circuit 9

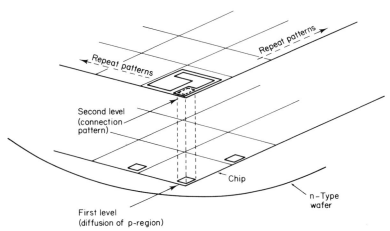

Fig. 1.3 Batch processing of *pn* junction diode illustrating the two patterning levels necessary.

of a thin layer of aluminum is insulated from most of the chip by a layer of silicon dioxide. Only where the actual *pn* junction is formed is the silicon dioxide not present so that:

a. The *p*-type impurities can be introduced or diffused into the *n*-type substrate to form the junction.
b. The aluminum can make electrical contact with the *p* region.

Thus, two pattern-formation steps can be identified. In the first, a rectangular "hole" has to be formed in the SiO_2 layer. When the *p*-type impurities are introduced, the insulating layer prevents them from entering the semiconductor in all places except where the SiO_2 layer has been removed. Thus, the *pn* junction is formed only in the localized region where it is required. This is illustrated in Figure 1.3 at the first level. Second, a pattern must be formed in the subsequent aluminum layer to define the shape of the electrical connection to the *p* region and to isolate it from the other parts of the circuit. This would, of course, be necessary if there were other circuit elements, such as resistors, capacitors, or transistors present. This pattern is illustrated in the second level shown in Figure 1.3. The second electrical contact, that to the *n* region, is made to the bottom face of the substrate and is often formed when the chip is mounted in its package. However, it is important to note that no pattern-formation stage is necessary in this instance. When each of these pattern-forming stages is carried out, they are carried out simultaneously by using an array of identical patterns. These patterns are shown for each level in Figure 1.3.

We can now summarize the processing of our simple circuit as follows (the chapter where each process is discussed in more depth is indicated):

a. Start with a "slice" or "wafer" (Chapter 2), of n-type silicon. [This is typically 30-100 mm ($1\frac{1}{4}$ to 4 in.) diameter and 0.5 mm (20 mils) thick].
b. Cover the surface with an insulating layer of SiO_2 ("oxidation," Chapter 4).

Level 1

c. Open holes in SiO_2 for formation of p region ("photolithography," Chapter 5).
d. Introduce p region impurities ("diffusion," Chapter 3).
e. Cover surface with aluminum ("metallization," Chapter 6).

Level 2

f. Define electrode pattern in aluminum ("photolithography").
g. Cover back surface with suitable alloy (e.g., gold antimony).
h. Carry out heat treatment to form low-resistance contacts between aluminum contact and the p region, and between the back contact and the n region.

Each of the processes a through h are performed on all circuits together on the wafer. Now the wafer is sawn or scribed and diced into chips, each containing one diode with its connections. All of the subsequent processing, such as mounting in its package and wire-bonding, must be carried out on each chip. This transition from batch to individual processing has an influence on the cost of the final circuit and is discussed earlier in this chapter. The dicing, mounting, wire-bonding, and hermetic sealing in a package of our final circuit are shown in Figure 1.4. A diamond saw or laser beam scriber is used to cut the silicon along tracks provided for at the mask stages. The die are mounted on headers by using a combination of alloying and ultrasonic vibration, and the top connection is made with a fine gold or aluminum wire between the contact on the chip and a pin connection in the package. Finally, a cover is put on providing a hermetic seal that protects the circuit from atmospheric conditions and physical damage. A detailed discussion of these techniques is given in Chapter 6.

In the chapters that follow we discuss in detail each of the process steps of integrated circuit fabrication.

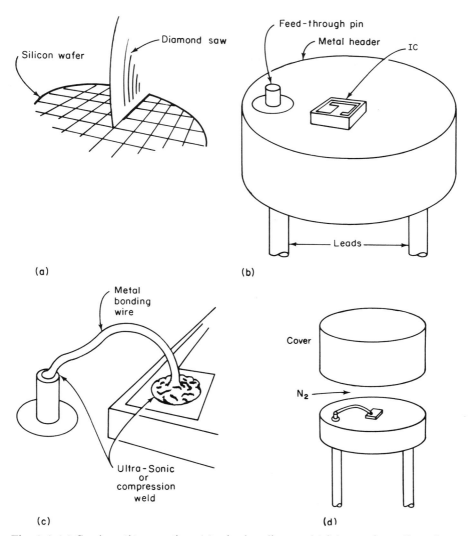

Fig. 1.4 (*a*) Sawing, (*b*) mounting, (*c*) wire-bonding, and (*d*) hermetic sealing of prototype integrated circuit.

Problems

1.1 The number of working circuits at the end of a process sequence is of particular concern to the IC manufacturer. If a manufacturer looses 5 percent of a lot at each stage and there are nine stages, determine the overall yield. $(.95)^9$

1.2 If the number of stages in problem 1.1 is doubled, by how many percentage points does the yield change?
1.3 Calculate the number of 0.25 in. by 0.25 in. square integrated circuits that can be made on a 4-in. diameter silicon wafer.
1.4 Calculate the number of 0.25 in. by 0.25 in. square integrated circuits that can be made on a 6 in. diameter silicon wafer.

CHAPTER 2

Materials for Semiconductor Devices

Instructional Objectives

This chapter presents an overview of some of the commonly used semiconductor material growth techniques. You will learn about:

a. Zone refining techniques.
b. Single crystal growth procedures.
c. Czochralski growth procedures.
d. Float zone crystal growth procedures.
e. Epitaxial crystal growth techniques.
f. Chemical etching techniques.

Self-Evaluation Questions

Watch for the answers to these questions as you read the chapter. They will help point out the important ideas presented.

a. Why is zone refining necessary in some base material?
b. How does Czochralski crystal growth differ from float zone crystal growth?
c. Why is float zone silicon usually cleaner than Czochralski?
d. Where is epitaxial growth material used?
e. What layer thicknesses are produced using vapor phase epitaxy?

f. How does MOCVD differ from VPE crystal growth?
g. At what temperature can plasma deposition take place?
h. How can chemical etching be used in making an IC?
i. How is a V-groove, as used in VMOS transistors, made?
j. Why will ion beam milling be a good technique in submicron structures?

2.1 Introduction

A key component in semiconductor microtechnology is the production and quality control of the base semiconducting material from which devices and integrated circuits are made. Semiconductor grade materials are usually composed of single crystals of high perfection and high purity, although amorphorus materials are also being investigated by many manufacturers. The production of this base material demands many years of concentrated research and development to achieve the necessary standard. The problem is clearly illustrated when one considers that William Shockley proposed a field effect transistor structure in 1952, but it was not until 1962 that silicon technology had developed to a point where such devices could be constructed with reproducible characteristics from one device to the next.[1] Today, silicon technology has reached the stage where such reproducibility and reliability is achieved in complex integrated circuits containing up to 100,000 transistors. This is quite an achievement! Gallium arsenide is another important semiconductor, but here the technology is much more primitive when compared to silicon. At the present time there is a concentrated effort to raise the standard of technology for gallium arsenide so that integrated circuits can be made that possess a very high operational speed resulting from the high electron mobility in this material. Each semiconducting material brings with it its own set of problems which must be solved before a useful device can be fabricated precisely, and with reproducible characteristics. Only then can factory production be contemplated. A further and important consideration that will decide whether a particular material or technique will be used commercially is the cost. To establish a new fabrication technique, it is first essential to show that improved performance will result, or alternatively that the new material (or device) has some feature unattainable in another material. Keeping these considerations in mind, we now outline some of the general features of present-day semiconductor

[1]W. Shockley, "A Unipolar Field Effect Transistor," *Proc IRE* **40** 1365–76 (1952).

technology. Our discussion is introductory, and, therefore, selective rather than exhaustive.

2.2 Crystal Purification and Growth

The majority of solids produced in nature are grossly imperfect and impure—a direct result of the haphazard way in which they are formed. Thus, the first stage in producing a semiconductor device is the purification of the material and then the growth of high-quality single crystals under controlled and ultra-clean conditions. Techniques of crystal growth have, in recent years, proliferated, each material having demanded its own variation to overcome its particular problems. We now briefly describe these techniques.

2.2.1 ZONE REFINING

Prior to the growth of a single crystal of silicon or germanium or gallium arsenide, the base material must be purified to a degree beyond that attainable by simple methods. For example, in some devices, the residual impurity content must not exceed about one impurity atom for every hundred million host atoms! A standard method of purifying material is zone refining, illustrated in Figure 2.1a. The general principle underlying this technique is that the uncounted impurity has a high segregation coefficient. Thus, if a rod of polycrystalline material is slowly passed through a heating coil, such that a molten zone passes from one end of the rod to the other, many impurities are concentrated in the liquid zone and are swept to the ends of the crystal (which are later discarded). If the refining process is repeated by multiple passes, then an extremely pure material can be prepared. An adaptation of this is the floating zone method. Here the rod is held vertically and a single molten zone is passed vertically from end to end. Surface tension in the liquid zone maintains the liquid between the two solid portions (Figure 2.1b). Since the solid acts as the container for its own melt, contamination from a crucible is avoided.

2.2.2 CRYSTAL GROWTH

At the end of the zone refining process, material of a high purity is obtained. The atoms in this solid, however, will not be arranged in the simple periodic structure that is characteristic of a single crystal (Figure 2.2). To change this material into a near perfect crystal, a process called

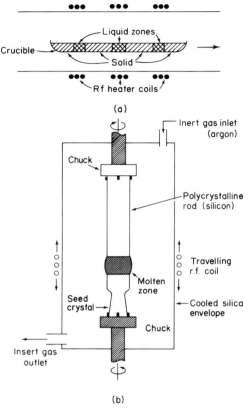

Fig. 2.1 (a) Zone refining of material using a horizontal boat; (b) zone refining and crystal growth using the boat-free vertical method.

crystal growth must be carried out. One way of achieving this is to convert the solid into a liquid and then to allow this liquid to solidify by moving it decisively into the correct position in the solid. This can be achieved by allowing the new solid to form on the surface of a "seed crystal." The seed is a small nucleus of crystal that in itself is highly perfect and, hence, can nucleate the structure of the new crystal. An alternative process relies on the conversion of the desired atom directly from the gas phase into a solid, this is termed vapor phase epitaxy (VPE). In crystal growth, the ideal situation is to have the seed of the same material as the growing crystal, as is always true with silicon. However, this is not always possible, particularly in some of the more recently developed binary (two-element), ternary (three-element), and quaternary (four-element) materials. We examine some of these processes.

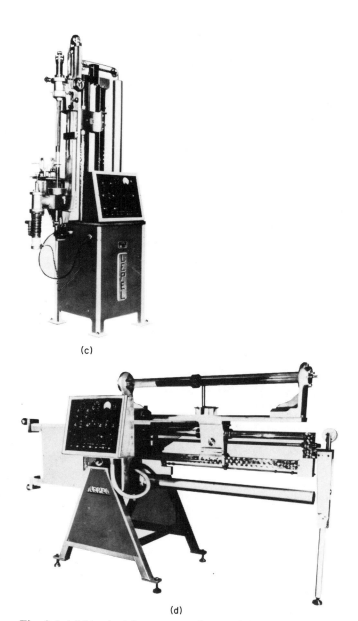

Fig. 2.1 (c) Vertical float zone refiner; (d) horizontal float zone refiner (Lepel Corporation).

18 Materials for Semiconductor Devices

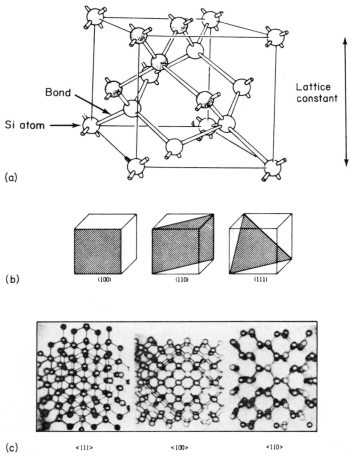

Fig. 2.2 The crystal structure of silicon: (*a*) the unit cell; (*b*) the three major crystallographic planes; (*c*) looking along the three major axes.

2.2.3 BULK GROWN CRYSTALS: THE CZOCHRALSKI GROWTH PROCESS

The growth of a bulk ingot by this process is shown schematically in Figure 2.3. The material for crystal growth is placed in a fused silica crucible and brought to a melt (typically around 1450° C), usually by means of a radio frequency (RF) heating coil. The seed is mounted on a chuck that can rotate about a vertical axis and may also be drawn slowly away from the melt, as is shown in Figure 2.3.

The procedure at this stage is to place the seed crystal at the surface of the melt and then draw it away slowly. As the seed is withdrawn, it

2.2 Crystal Purification and Growth   19

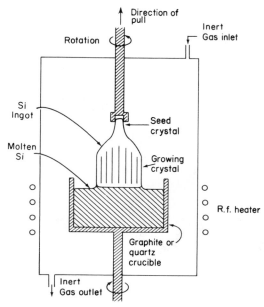

Fig. 2.3 The Czochralski crystal growth process.

raises a film of liquid that cools, solidifies, and is forced to take on the structure of the seed. The pull is continued smoothly, thus allowing a cylindrical ingot to grow, as is illustrated in the photograph shown in Figure 2.4a, which shows an 80-mm (3.1-in.) diameter gallium arsenide ingot. In the early days of semiconductor processing technology, only relatively small diameter ingots (30 mm, 1.25 in.) could be drawn; in contrast with the large diameter ingots (150 mm, 6 in.) that are readily obtained with current technology. An example of this type of crystal puller is illustrated in Figure 2.4b, which shows the "Melbourn" III-V crystal puller. Note that in this instrument, the entire growth process is carried out at low pressures and high temperature and the whole system has to withstand these extreme conditions. Another very important practical point relates to the diameter of the initial section of the growing crystal. The diameter of the growing crystal is encouraged to decrease initially and then to increase over the first centimeter ($\frac{1}{2}$ in.). This is achieved by starting the pull at a relatively high speed; and then slowing down the pull rate and the growth rate determines the diameter of the growing ingot. This necking technique allows any dislocation to be terminated in the necked region, resulting in a much higher quality crystal.

The process described above will ideally produce an intrinsic

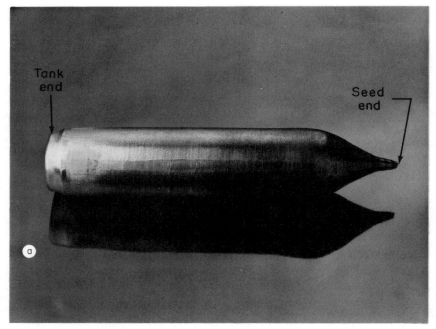

Fig. 2.4 (a) A single crystal of gallium arsenide (8-cm diameter) grown by the Czochralski process.

semiconductor[2]. Alternatively, material may be doped p- or n-type by introducing the appropriate p- or n-type doping pellet into the melt. It should be stressed that the growth of a binary solid such as gallium arsenide (GaAs) is much more complex than, for example, silicon. The details of this process are the domain of a more advanced book.

2.2.4 FLOAT ZONE PROCESS

This technique is in essence an extension of the zone refining described previously and illustrated in Figure 2.1a and b; the difference here being that a seed crystal is clamped on to one end of the polycrystalline rod. The position of the RF heating coil is such as to produce a liquid zone at one end of the polycrystalline rod, which is supported by surface tension forces. A seed crystal is placed into the liquid zone. The liquid zone is now moved along the rod, allowing the rear surface to cool and solidify in the structure of the seed crystal. To dope the crystal, the dopant can either be distributed uniformly in the polycrystalline rod, or

---

[2]An intrinsic semiconductor is one that contains no intentionally induced impurities.

2.2 Crystal Purification and Growth

Fig. 2.4 (*b*) The "Melbourn" puller used to grow the gallium arsenide crystal (permission of Cambridge Instruments Ltd, England).

alternatively it can be inserted periodically into the rod by drilling a sequence of holes along the length and loading them with the dopant material. For the best uniformity, however, certain dopant materials may be included in the surrounding gas. The gas phosphine, for example, can provide the phosphorus needed for n-type doping of silicon. One of the most important advantages of this technique over the Czochralski process

22 Materials for Semiconductor Devices

Fig. 2.4 (c) Float zone refiner (Lepel Corporation); (d) Czochralski crystal puller (NRC Corporation).

is that the molten silicon is not contained in a fused silica crucible and it is thus less susceptible to contamination by unwanted impurities, particularly oxygen. Crystals with very high values of resistivity have been grown using this technique.

2.3 Epitaxial Growth

Bulk grown crystals of a quality sufficiently good for direct device fabrication are difficult to grow. One way of circumventing this difficulty is to use a polished slice of bulk grown crystal as a foundation and to grow onto this surface a much higher grade epitaxial layer of semiconductor material for the actual device structures.

In the process of epitaxial growth, a polished bulk grown sample (i.e., a wafer) is used as the seed for epitaxial growth. The word epitaxial means that the new crystal layer has the same crystal axes as the base wafer. The detailed processes of epitaxial growth are numerous and varied according to the needs of the different types of semiconductor. A simple crystal, such as silicon, clearly will differ in its needs from a more complex compound semiconductor, such as gallium arsenide.

2.3.1 LIQUID PHASE EPITAXY (LPE)

The basic principle underlying the process of liquid phase epitaxy is shown in Figure 2.5a. A substrate (seed) crystal is held above a semiconductor melt and then dipped into it. As the substrate is withdrawn from the melt to the cooler region of the furnace, the molten film covering the surface will form an epitaxial crystalline layer provided that the rate of cooling (dependent on the withdrawal rate) is carefully controlled.

Practical systems, although more complex, rely on the same basic principle. For example, Figure 2.5b shows the apparatus used for the liquid phase epitaxy of the semiconductor material gallium arsenide. In this sophisticated system, the semiconductor growth is held in a graphite boat located in a fused silica tube with very pure hydrogen gas flowing through the system. Also shown in this figure is a cross section of the graphite boat and its melts with the gallium arsenide source crystal at the bottom and top of the melts. The bottom section is a slider that is used to slide the bottom source crystal or the growth substrate into position under the melt. The pure gallium melt dissolves increasing amounts of arsenic from the gallium arsenide source crystal as its temperature is raised. Once in the gallium melt, the arsenic can be removed by cooling the melt in contact with the substrate single crystal (i.e., by sliding the crystal to be used for growth under the melt). This forces gallium arsenide to grow on the substrate surface. Variations of this technique may be used for other crystal systems.

2.3.2 VAPOR PHASE EPITAXY (VPE)

Like liquid phase epitaxy, there are many variations of the basic technique of VPE. Figure 2.6 illustrates the principles of a silicon growth scheme. A substrate (single crystal) is placed inside fused silica furnace tube and held at approximately 1250° C. Silicon tetrachloride vapor carried in a stream of hydrogen gas is passed through the furnace. Inside the heated region the chemical reaction

$$SiCl_4 + 2H_2 \rightarrow Si + 4HCl \qquad (2.1)$$

takes place. The silicon produced in this way is deposited and forms a single crystal on the substrate surface. Very pure epitaxial films can be

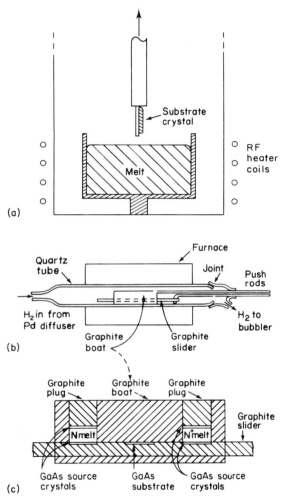

Fig. 2.5 (a) A schematic of the liquid phase epitaxial (LPE) growth system; (b) an LPE system for gallium arsenide; (c) an enlargement of the boat assembly (permission of Cornell University).

fabricated by controlling the purity of the chemicals; alternatively, the crystal can be deliberately doped n- or p-type by first bubbling the hydrogen through a weak solution of phosphorous trichloride (n-type) or boron trichloride (p-type). This procedure will produce epitaxial layers to 2 to 20 μm thick with a resistivity that is controllable to within 5 percent from one crystal to the next.

Another example of a VPE system, used for growing compound semiconductors such as GaAs and InP, is shown in Figure 2.7. Consider the

2.3 Epitaxial Growth

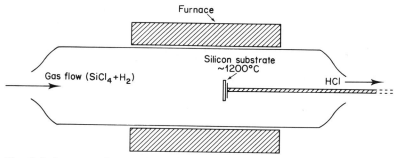

Fig. 2.6 A vapor-phase epitaxial (VPE) system for silicon.

growth of GaAs. Gallium is used as the source material and is held at 800° C. A crust must form over the entire surface before crystal growth begins to ensure arsenic saturation. Before crystal growth begins, a back etch (removing) of the surface is carried out by running the system at the relatively high temperature of 850° to 900° C. Thus assured of a good growth surface, the epitaxial layer is deposited at 710° to 750° C. Highly doped material can be grown by moving the material upstream and can be lightly doped by moving it downstream in the H_2S gas. By balancing the arsenic trichloride flow relative to the H_2 flow, it is possible to control the ambient net donor density over a significant range, including the growth of very pure undoped (or intrinsic) materials.

To conclude our discussion of crystal growth, we describe two relative newcomers to the process, molecular beam expitaxy (MBE) and metal organic chemical vapor deposition (MOCVD).

2.3.3 MOLECULAR BEAM EPITAXY (MBE)

The basis of this technique is to allow a beam of the desired constituent atoms to fall on to, and stick to, a desired substrate held at an elevated

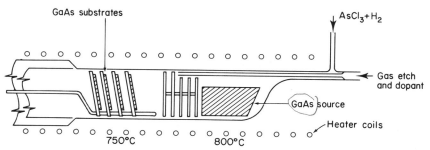

Fig. 2.7 A VPE system for gallium arsenide (permission of Plessey Company).

temperature in an ultra-high vacuum, which is chosen to allow the best quality crystal to grow on the substrate. In the case of compound semiconductors, separate cells for each of the component elements provide the atoms required for growth. The sources of the molecular beams, called Knudsen effusion cells, are in essence heated enclosures containing the elements required for the molecular beam. The elevated temperature ensures the desired high vapor pressure, and a suitable orifice allows the beam to emerge in the desired direction. In practice, the difficulty in growing a multi-atom compound semiconductor, such as gallium-indium-arsenide (GaInAs), is to ensure that each of the sources is at the correct temperature to allow the molecular beams arising from the vapor pressure in the cell to form a stoichiometric solid.

A schematic diagram of a typical molecular beam epitaxial system is shown in Figure 2.8. Basically, this is a very expensive technique, since all the crystal growth takes place in an ultra-high vacuum chamber ($\cong 10^{-9}$ Torrs). This, however, also has its advantages, since a range of auxiliary facilities can be placed inside the chamber. An electron gun allows the electron diffraction patterns of the crystal to be monitored throughout the growth, and the concentration of the molecular beam constituents may be monitored precisely by using a quadrupole mass spectrometer. The doping of epitaxial layers is not difficult provided that the dopant atoms have high sticking coefficients to the epitaxial layer concerned (i.e., that they have a high probability of sticking to the growth surface).

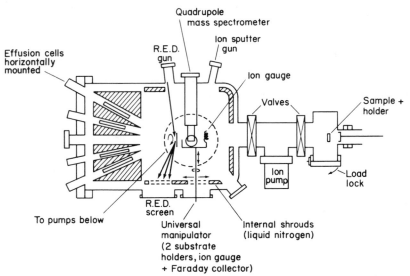

Fig. 2.8 A schematic layout of an MBE system (permission of Cornell University).

2.3.4 METAL-ORGANIC CHEMICAL VAPOR DEPOSITION (MOCVD)

This process is also known as organometallic chemical vapor deposition (OMCVD). Its basic principles are shown in the schematic diagram in Figure 2.9a and b. To illustrate these basic principles, we consider the growth of gallium arsenide; a group three element Ga (in the form of trimethyl gallium) is transported with the H_2 gas to the reactor. Similarly, the group five element As (in the form of arsine) is similarly transferred and meets the gallium component just before the entrance to a cold walled reactor. Inside the reactor, the two gasses pass over an inductively heated

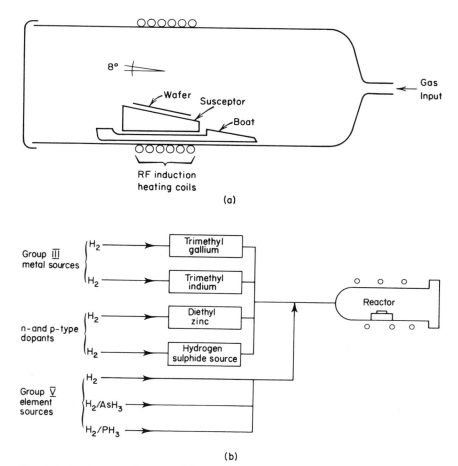

Fig. 2.9 A schematic layout of the gas input lines for metal-organic CVD (MOCVD) of gallium arsenide (permission of British Telecom).

substrate where the reaction

$$Ga(CH_3)_3 \uparrow + AsH_3 \uparrow \xrightarrow[\text{substrate}]{\text{hot}} GaAs(solid) + 3CH_4 \uparrow \quad (2.2)$$

occurs only on the heated substrate. A temperature range of 600 to 800°C is typical of those required for the reaction. If doped semiconductors are required, then p or n-type dopants can be introduced into the gas in the manner shown in Figure 2.9. This technique differs from vapor phase epitaxy in that the growth reactor vessel is cold; only the crystal seed wafer is heated to initiate the crystal growth.

The latter two techniques have much in common. They are both best suited to the growth of special profiles, which are not readily obtained by the other established technologies. Such profiles are used, for example, in microwave and optoelectronic-device structures. The two techniques are essentially of a nonequilibrium nature and can be used to achieve very rapid changes in doping profile. Both methods are being studied in many laboratories, and the current achievements are very encouraging for future applications.

2.3.5 CHEMICAL VAPOR DEPOSITION

Device fabrication processing frequently requires the deposition of thin insulating layers onto a semiconductor surface. This can be achieved by sputtering (described earlier) or by thermal oxidation, as in the special case of silicon (Chapter 4). Another important class of deposition is chemical vapor deposition (CVD), which is regularly used for silicon dioxide, silicon nitride, and polycrystalline silicon (polysilicon). The basic principle of the technique is to create a reaction in a material at a high temperature that enables it to be deposited on a crystal surface with the aid of a carrier gas. The vapor phase epitaxial technique is a good example of this process.

For silicon dioxide, deposition can be achieved via the reaction of silane at 200° to 500° C using an N_2 carrier gas

$$\text{Silane } (Si(H_4)) + 2O_2 \rightarrow SiO_2 + 2H_2O \quad (2.3)$$

for silicon nitride at 750° to 850° C

$$3SiH_4 + 4NH_3 \rightarrow Si_3N_4 + 12H_2 \quad (2.4)$$

and for polysilicon at 600° to 850° C

$$SiH_4 \rightarrow Si + 2H_2 \quad (2.5)$$

This last reaction can be used for epitaxial growth and, as we outlined previously, can be used to produce a range of doped material.

Polysilicon is frequently used in MOS integrated circuits where it is valuable as an additional conducting layer when it is heavily doped with boron (p-type) or phosphorus (n-type) to render it low resistance. A further discussion on deposited films is presented in Chapter 4.

2.3.6 PLASMA DEPOSITION

One of the main difficulties with the use of CVD techniques is the relatively high temperatures involved. At the lower deposition temperatures the layer quality, particularly in insulators, is not good enough for device fabrication. Furthermore, the relatively high temperatures cannot be tolerated when metallizations (Chapter 6) such as aluminum are present on the surface.

In plasma deposition, the active species is present in a gas plasma, enabling reactions to take place at around room temperature, which would not otherwise occur outside the plasma. This technique is being used to produce good quality silicon nitride, which is very valuable in silicon and gallium arsenide technology. The semiconductor wafer is placed inside the plasma vessel and the (gaseous) active species is allowed to react and deposit on the surface.

2.3.7 WAFER PREPARATION

In order to fabricate devices or integrated circuits, the starting material, the bulk crystal, must first be sliced into wafers. Devices may then either be fabricated directly in the bulk crystal, or indirectly by using the bulk crystal as a seed for epitaxial growth, as was discussed earlier. In many applications the bulk crystal rod is ground into a cylinder shape to ensure a consistency in the diameters of the eventual wafers—this is important for production line handling in subsequent IC processing. Automation does depend critically on handling wafers with identical dimensions. To ensure this situation the boule is ground on a lathe to form a constant diameter cylinder. The orientation of the surfaces normal to the cylinder axis is fixed by the seed crystal (and for silicon, gallium arsenide, and indium phosphide it is usually $\langle 100 \rangle$ and occasionally $\langle 111 \rangle$).[3] The orientations within the plane must be fixed by X-ray diffraction. Once the geometry of the cylinder is determined, a flat is ground along its length to provide a permanent reference on each of the final wafers

---

[3]The symbol $\langle \ \rangle$ denotes Miller indices. See Appendix 6, reference 6 for further details.

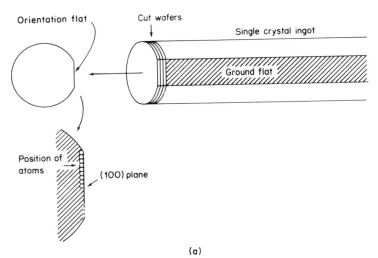

(a)

Fig. 2.10 (*a*) Illustrating a single ingot of silicon with the ground flat aligned to a (100) plane.

Fig. 2.10 (*b*) Commercial polishing/lapping machines (Strasbaugh Model No. 46UR-1).

(Figure 2.10). This step is very important, since the final crystal wafer will have preferential cleavage planes which can be used for breaking the whole wafer into its component devices or ICs. The edges of the separate ICs must be aligned so that they are parallel to the cleavage planes. The cylindrical rod is then sliced into thin parallel wafers, as is shown in Figure 2.10. For this delicate cutting process, a special diamond-impregnated wheel is used (Figure 2.11). A fine ring-shaped blade is coated with diamond powder (i.e., very fine particles of diamond), which is sufficiently hard to saw away the crystal rod.

Two forms of diamond saw may be used, one with the cutting edge as the outer surface or, alternatively, one with the cutting edge on the inside. These are illustrated in Figure 2.11. The mechanical rigidity of

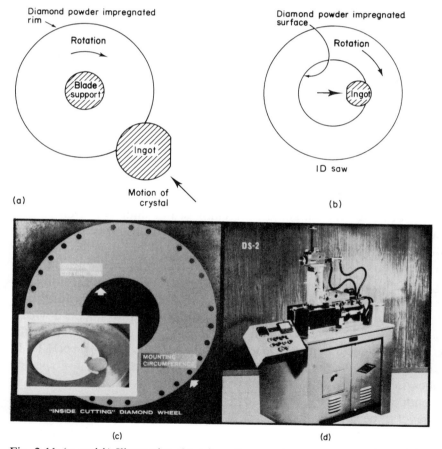

Fig. 2.11 (*a and b*) Illustrating the microslice annular saw process; (*c and d*) photographs of a real saw (Hamco Division of Kayex Corporation).

the saw with the internal cutting edge enables the use of a thinner blade wheel that does not ripple and, hence, produces a thinner cut with less wasted semiconductor material between slices.

At this point in the procedure, the sawing process leaves wafers with some marks that must be removed by using a suitable etchant (i.e., a chemical solution that will slowly dissolve the surface region of the semiconductor). This chemical polishing process will remove the damage from the immediate cutting area at the surface of the wafer. Unfortunately, the cutting damage will propagate some distance into the slice and must be removed if a surface is to be presented for crystal growth or for direct device fabrication. This cannot be done by continued etching, primarily because etching will increase rather than decrease surface topographical irregularities. To avoid this difficulty, a process of mechanical/chemical polishing is used. The wafer is mounted on a rigid arm suspended over a rotary polishing disk. Polishing takes place by impregnating the disk with a chemical etchant that will slowly dissolve the semiconductor surface. This technique ensures that the wafer surface is polished to an optically flat finish. As a final step, the wafer is given a short conventional chemical etch to remove any residual damage and is cleaned to remove any residual contamination making it ready for device processing or alternatively as the substrate (seed crystal) in an epitaxial growth process.

2.4 Chemical Etching

In the preceding section, chemical etching or polishing is presented as a means of removing surface damage resulting from the sawing of the wafers. But surface etching is used in other ways in microtechnology. One important use is in the successive diffusion (or ion implantation) stages of making an IC. Here it is necessary to cut diffusion windows in the protective surface oxide, to remove the oxide completely, or to cut a window in the oxide for metallization purposes. For silicon devices, the oxide used is thermal silicon dioxide (SiO_2). A suitable etchant for this is hydrofluoric acid (HF) or a mixture of this acid with NH_4F. In addition, a number of other etches have been developed for specific applications. Some of the more common solutions are summarized in Appendixes 2 and 3. Etches are also needed to selectively remove metal overlayers; the specific etches depend on the metal that is to be removed.

Returning to the crystal surface itself, we must point out that a number of specialty etches have been developed over the years. Some etches produce mesa structures; other etches are used for selectively decorating dislocations to determine the perfection of the starting material; and some etches are used for (delineating) *pn* junctions. In this last category, chemicals that act differently on *p*- and *n*-type material are used. By selectively etching on one side of a junction, its actual position can be made visible and measured (Appendix 1). In the second category, the etch acts more rapidly on the less ordered material in the vicinity of a dislocation and, hence, produces a small pit at the point where the dislocation reaches the surface.

2.4.1 ANISOTROPIC ETCHES

Recently, several etches have evolved that are capable of selectively producing a trough with either a V- or a U-shaped cross section. These etches are called anisotropic etches, and their successful use in device fabrication reflects the advanced state of silicon technology.[4] To illustrate the principles of this technique, let us consider one V-groove etching technique. The basis of operation of these etches is related to the variation in packing density of the different planes in the silicon (or gallium arsenide) crystal structure. Figure 2.2 illustrates this point and defines the three major planes $\langle 100 \rangle$, $\langle 110 \rangle$, and $\langle 111 \rangle$ in the crystal. From this figure, it is readily seen that the atomic packing density and available bonds decrease as we go from $\langle 111 \rangle$ to $\langle 100 \rangle$ to $\langle 110 \rangle$ and, hence, selective etches are able to remove atoms in some directions more readily than others.

This principle is illustrated in Figure 2.12, which shows an oxide-coated $\langle 100 \rangle$ silicon slice with two long window cuts parallel to the $\langle 110 \rangle$ directions. When a sample is placed in an orientation dependent etch, for instance KOH, normal propanol, and H_2O, the etching will proceed to the $\langle 100 \rangle$ direction until the etch front hits the two $\langle 100 \rangle$ planes intersecting the $\langle 100 \rangle$ plane at the edge of the mask opening; then, etching will stop with the resulting V-groove. The etch depth a to oxide opening L (ratio a/L) is approximately 1.707 and is fixed by the photolithographic opening L and the geometry of the crystal structure.

As an example of the application of an anisotropic etch, Figure 2.15 shows the structure of a V-groove channel FET in silicon.

2.4.2 PLASMA ETCHING

The wet chemical etching processes described above have limitations when very fine patterns have to be produced. This is due to the viscosity limit of the liquids needed to bring in new active radicals and to the removal of the reaction products of the etch. These limitations can be

---

[4]K. E. Bean, "Anisotropic Etching of Silicon" IEEE Transactions on Electron Devices, ED-25 no. 10, October 1981, pp. 1185–1193.

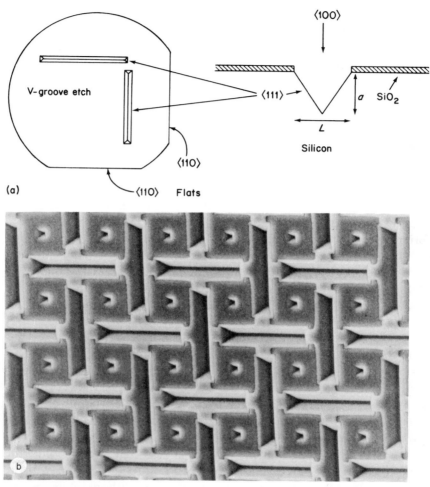

Fig. 2.12 (*a*) Illustrating the V-groove anisotropic etching process; (*b*) V-grooves in silicon (permission of Siliconics Ltd, Swansea).

overcome by plasma etching. In this process, the etching is achieved by using the active species found in a plasma to attack the semiconductor surface.

The earliest plasma process used oxygen to remove photoresist. As an example of a plasma suitable for etching silicon, silicon oxide, or silicon nitride, the complex reactions resulting from CF_4 gas can be used. The details of practical systems are complex, and interested readers should refer to Appendix 6, reference 5 for further details.

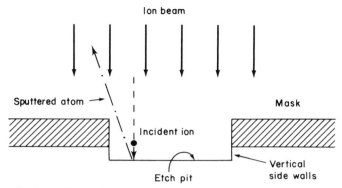

Fig. 2.13 Illustrating the ion milling process.

2.4.3 ION BEAM ETCHING (ION MILLING)

Ion milling is in essence the process whereby the surface of a solid is slowly eroded by using an ion beam. A collimated beam of ions with energies in the range 500 eV to 1 keV (Figure 2.13) is directed onto the surface of the solid. These ions can strike the near-surface atoms of the solid and, in the process, communicate to them sufficient energy to enable them to leave the surface. This process is called sputtering. The use of argon ions as the sputtering beam is a convenient working gas. One important advantage of this technique is that it produces nearly vertical etch pits with no problems of undercutting at the edges. This is important for future developments of submicron structures.

Problems

2.1 What is meant by the term "single crystal"?
2.2 How does MOCVD differ from VPE?
2.3 What is the atomic spacing between two silicon atoms?
2.4 What is the atomic spacing between two germanium atoms?
2.5 What is the atomic spacing between two gallium arsenide atoms?
2.6 What is meant by the term "polysilicon"?
2.7 Why must a semiconductor wafer be etched after wafer cutting?
2.8 What type of etches are used to produce V- or U-shaped troughs in silicon?

CHAPTER 3

Impurity Doping in Semiconductors

Instructional Objectives

This chapter introduces the concept of intrinsic and extrinsic semiconductors. After reading this chapter you will be able to:

a. Explain intrinsic semiconductors.
b. Explain extrinsic semiconductors.
c. Describe majority carriers in semiconductors.
d. Describe minority carriers in semiconductors.
e. Describe doping techniques.

Self-Evaluation Questions

Watch for the answers to these questions as you read the chapter. They will help point out the important ideas presented.

a. Why is the intrinsic concentration for most semiconductors not zero at 300 K?
b. Why does silicon become *p*-type when boron is introduced as an impurity?
c. Why are impurities such as copper troublesome in some semiconductors?
d. What are the two most common methods used to introduce impurities into semiconductors?

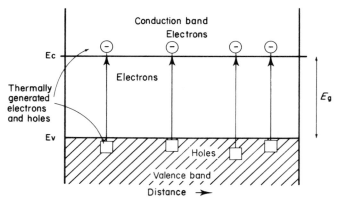

Fig. 3.1 A simple band diagram of a semiconductor.

3.1 Basic Concepts

3.1.1 INTRINSIC SEMICONDUCTORS

Semiconducting material may, for convenience, be divided into two categories: intrinsic and extrinsic.[1] We consider first the intrinsic semiconductors. In their pure form, these materials are insulators at very low temperatures but begin to conduct electricity as they are heated by virtue of their relatively narrow band gaps. Figure 3.1 shows an energy band diagram of an intrinsic semiconductor. There are, in fact, no fundamental differences between intrinsic semiconductors and insulators; the division is somewhat artificial since, by convention, intrinsic semiconductors are defined as insulators with band gaps of less than about 2 ev. The reason for this division becomes clear if one considers the statistical process taking place within solids when thermal energy (heat) is supplies to the crystalline atoms. When this energy is communicated to the valence electrons, some will (statistically) gain sufficient energy to cross the forbidden gap and produce both electrons n and holes p, which may then move under the influence of an applied electric field to produce current. A full mathematical treatment of the problem establishes that the number of electron-hole pairs at temperature T (expressed in degrees Kelvin) is given by the equation

$$p = n = n_i = G \exp\left(-\frac{Eg}{2kT}\right) \qquad (3.1)$$

[1]For further discussion, see Appendix 6, reference 6.

TABLE 3.1

Semiconductor	Ge	Si	GaAs
Eg (eV)	0.67	1.11	1.40
n_i (300 K)cm^{-3}	2.4×10^{13}	1.45×10^{10}	1.1×10^7

where G is a constant that varies from material to material.[2] In Table 3.1, the value for n_i at 300 K for some important semiconductors is given. From this table, we see that at room temperature n_i varies from approximately 10^{13} cm^{-3} for germanium to $\cong 10^7$ cm^{-3} in gallium arsenide. Thus, the number of electrons and holes that are available to transport electricity decreases by six orders of magnitude for the relatively small change in band gap from 0.67 eV to 1.40 eV. As a result of the strong exponential dependence of n_i on band gap Eg, a semiconductor with a band gap of greater than about 2 eV is considered to be an insulator at room temperature because of its low value of n_i and, hence, its low conductivity.

Note that the conductivity σ is given by

$$\sigma = nq\mu_n + pq\mu_p$$
$$\sigma = n_i q(\mu_n + \mu_p) \quad (3.2)$$
$$\sigma = q(\mu_n + \mu_p)G \exp\left(-\frac{Eg}{2kT}\right)$$

hence

$$\log \sigma = \ln (\text{constant}) - \frac{Eg}{2k}\left(\frac{1}{T}\right) \quad (3.3)$$

Here it is assumed that the temperature dependence of G and the mobilities ($\mu_n + \mu_p$) are negligible.

Thus, a plot of log σ versus ($1/T$) will be a straight line (Figure 3.2) with a slope of $-(Eg/2k)$. This is a valuable experimental method of determining the band gap Eg. This type of plot is commonly known at an Arrhenius plot.

3.1.2 EXTRINSIC SEMICONDUCTORS

An *extrinsic* semiconductor is so named because its conductivity is produced by the *external* influence of specific impurity atoms. Atoms with a valence of either three or five are substituted in place of regular

---
[2]See Appendix 6, reference 6.

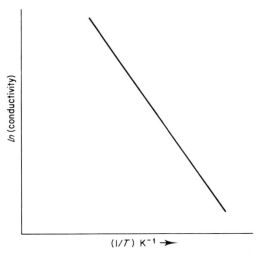

Fig. 3.2 A plot of *ln* (conductivity) versus (1/T) for an intrinsic semiconductor.

silicon or germanium atoms. It is important to stress that these impurities must sit on a regular silicon (or germanium) site if they are to be electrically active (Figure 3.3a and b) and not on an interstitial site, as is shown in Figure 3.3c.

Impurities atoms such as phosphorus or arsenic, which have five valence electrons, are termed donor atoms since they have one more electron than is required for the covalent bonding. This excess electron can easily be pulled away from its parent atom, allowing it to respond freely to an electric field. This is reflected by the relatively small electron binding energy ΔE (required to free the electron), that is to raise it from its present level to the conduction band. The electron binding energy ΔE

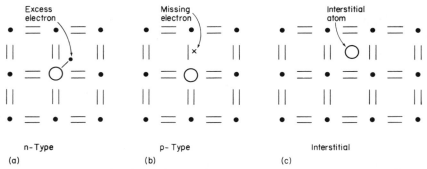

Fig. 3.3 A schematic picture illustrating (*a*) donor, (*b*) acceptor, and (*c*) interstitial atoms in silicon.

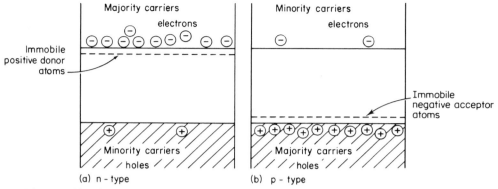

Fig. 3.4 Simple band diagrams of (a) *n*-type and (b) *p*-type silicon.

is much less than Eg, and is typically of the order of 0.05 eV for practical donor atoms. The band picture corresponding to this situation is shown in Figure 3.4a. This type of material is called *n*-type because the conductivity is carried by the negative charge carriers (electrons).

The alternative situation resulting from the introduction of an impurity with three valence electrons, such as boron or aluminum in silicon, is shown in Figure 3.4b. In this situation, the covalent bonding scheme is deficient of one electron and, if this atom can attract (i.e., accept) an electron from one of its neighbor atoms, it will produce a "hole" and result in conductivity. These are called acceptor impurities. In this circumstance, the conductivity will be by the positive holes, and the material is termed *p*-type.

If an *n*-type semiconductor is doped with N_D atoms/cm^{-3} where $N_D \gg n_i$, then there will be N_D electrons in the conduction band (i.e., $n \approx N_D$) and the conductivity will be

$$\sigma_n = nq\mu_n + pq\mu_p$$
$$\cong N_D q\mu_n + 0$$

The number of holes p can be ignored for the following reasons. As the number of electrons increases above n_i, there will be an increase in the probability of an electron and a hole colliding and annihilating each other. The technical word for this is *recombination*. As a result of this process, when *n*-type doping is introduced, n increases and p decreases. A formal mathematical treatment of this phenomenon results in the *law of mass action*:

$$pn = n_i^2 \tag{3.4}$$

since
$$n = N_D \qquad (3.5)$$
$$p = n_i^2/N_D$$

Consider the case of n-type silicon at 300 K with $n_i \cong 1.45 \times 10^{10}$ cm^{-3}. A typical value of doping for a semiconductor device application might be 10^{18} cm^{-3}, and

$$p \cong \frac{n_i^2}{N_D} = \frac{(1.45 \times 10^{10})^2}{1 \times 10^{18}} \text{ cm}^{-3}$$
$$p \cong 2.10 \times 10^{+2} \text{ cm}^{-3}$$

justifying the assumption that $p \ll n$ and can be ignored.

The same arguments may be used for p-type material:

$$\sigma p \cong pq\mu_p$$

where
$$n = n_i^2/N_A \ll p \qquad (3.6)$$

As a result of the small values of ΔE for these donor and acceptor atoms, it only requires a small quantity of thermal energy to take away the donor electrons (n-type) or to attract a valence electron (p-type). Consequently, these centers become electrically active at low temperatures, typically 20 to 40 K. Conversely, these dopants become electrically inactive at $T = 0$ K. The conductivity versus temperature will take the form shown in Figure 3.5 where T_F is the freeze out temperature—so called because the carriers are frozen in their donor (or acceptor) atoms below this temperature. The conductivity rises abruptly at T_F and to a first approximation remains constant at a value of $\sigma = N_D q \mu_n$ for n-type ($\sigma = N_A q \mu_p$ for p-type). This is only an approximation, since the mobility is a function of temperature and decreases slowly as the temperature is increased.

A very important point that must not be forgotten is the constant presence of intrinsic carriers n_i resulting from the finite temperature of the semiconductor. The relation n_i is strongly temperature dependent, and at some elevated temperature T_i, n_i will exceed N_A (or N_D). At this temperature, the semiconductor will revert to being intrinsic and this represents an upper temperature limit for the operation of semiconductor devices. Note also that for a given semiconductor, T_i will be dependent on the doping density N_D (or N_A).

Up to this point, our discussion has been concerned with only one

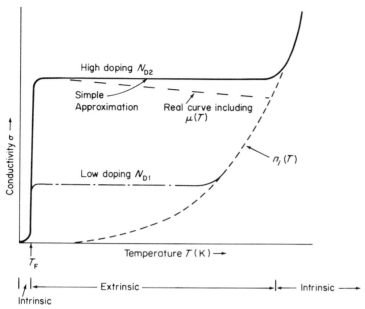

Fig. 3.5 A schematic plot of conductivity versus temperature (K) for an extrinsic semiconductor.

type of dopant: p-type or n-type. It is interesting to speculate, however, on what may happen if the material contains a mixture of p-type and n-type dopants. Let us assume that N_D is greater than N_A. The donors will donate all their electrons to the conduction band, and the acceptor atoms will donate all their holes to the valence band. In these circumstances, the increase in both electron and holes will result in enhanced recombination, thus removing both electrons and holes in the process. A simple explanation of the situation would be that some of the electrons from the donors have passed indirectly onto the acceptor atom levels. As a result of this process, the net number of electrons is $n = (N_D - N_A)$ and not the sum of the two components $N_T = (N_D + N_A)$. Generalizing this process, we can identify the following possibilities:

a. $N_D > N_A$ and $(N_D - N_A)$ gives a net electron combination—n-type.
b. $N_A > N_D$ and $(N_D - N_A)$ gives a net hole concentration—p-type.
c. $N_A = N_D$ and $(N_D - N_A)$ is zero, and the material appears to be intrinsic.

Case c is very important because all of the external carriers neutralize each other, and the only carriers remaining in the semiconductor will be the intrinsic electrons and holes (i.e., $n_i = n = p$). In these circumstances, the material is said to be *compensated*. This turns out to be a very practical way of producing a quasi-intrinsic semiconductor. For example, if after an extensive purification of a solid (Chapter 2) the residual impurities are p-type with a given concentration of N_A cm^{-3}, then the crystal can be regrown with exactly N_D ($=N_A$) donors added intentionally to the lattice. The semiconductor is compensated and appears to be intrinsic. In a material like GaAs with $n_i \cong 1.1 \times 10^7$ cm^{-3}, true intrinsic material is not available. The current practice is to grow the best material that technology can achieve (which is a net donor density of around 5×10^{14} cm^{-3}, seven orders of magnitude too large) and to compensate with chromium.

At this point it is important to highlight another problem related to doping. If a semiconductor contains N_D donors and N_A acceptors, where for argument we let $N_A > N_D$, the net carrier density will be

$$n = (N_A - N_D) \tag{3.7}$$

and the conductivity will be

$$\sigma = (N_A - N_D)q\mu_p \tag{3.8}$$

Consider now the microscopic origin of the mobility μ. The mobility μ is defined as the mean drift velocity of the electrons (or holes) per unit electric field.

If the mean drift velocity of the electrons due to the electric field F is u, then

$$\mu = \frac{u}{F}$$

The electrons (or holes) are at all times undergoing a random walk process (Figure 3.6) and are undergoing frequent collisions with atoms at a mean distance (free path) λ with a mean time between collisions τ. A rigorous treatment of the problem shows that the mobility

$$\mu = \frac{q\tau}{m^*} \tag{3.9}$$

where m^* is the effective mass of the charge carrier.[3] Now it must be remembered that any donor or acceptor atom will be charged either negative (acceptor atoms) or positive (donor atoms), and that they provide

---

[3]See Appendix 6, reference 6, for further details.

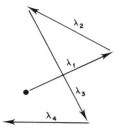

Fig. 3.6 A schematic diagram showing the random motion of a charge carrier (electron or hole) in a semiconductor.

effective centers for scattering electrons and holes. Consequently, large numbers of donors and acceptors will both decrease λ and τ and, hence, decrease the mobility μ. In this situation, it is the total number of scattering centers $(N_A + N_D) = N_T$ that controls the mobility μ and not the difference, as in the conductivity. An important consequence of this, for example, is that a compensated semiconductor will appear to be intrinsic if its conductivity is measured, but can have a mobility much lower than a true intrinsic semiconductor.

This is illustrated for silicon in Figures 3.7 and 3.8. Figure 3.7 shows the relationship between μ_n and μ_p and doping density N_A or N_D. Figure 3.8 shows how $\sigma_n = N_D q \mu_n (N_D)$ and $\sigma_p = N_A q \mu_p (N_A)$ vary and shows N_D and N_A in circumstances where there is only one type of carrier N_D or N_A.

In the more general case, the following procedure must be adopted:

a. N_T can be used to determine μ through Figure 3.7.
b. $(N_D - N_A)$ yield the net carrier density n_c and, hence,

$$\sigma = n_c q \mu (N_T)$$

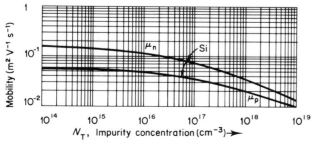

Fig. 3.7 A plot of mobility μ_n and μ_p versus total impurity concentration N_T for silicon [after Sze (reference 6, Appendix 6)].

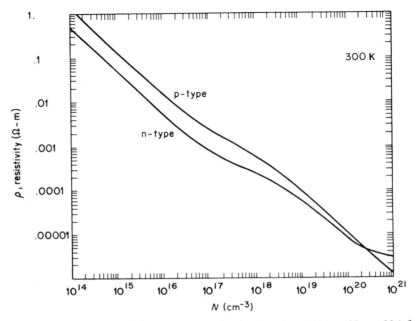

Fig. 3.8 A plot of resistivity ρ versus doping density N ($N = N_A$ or N_D) for n- and p-type silicon.

We can now clarify these relationships by determining the mobility and conductivities of two silicon crystals:

a. With $N_D = 5 \times 10^{15}$ and $N_A = 1 \times 10^{15}$ cm^{-3}
b. With $N_D = 2.5 \times 10^{16}$ and $N_A = 2.1 \times 10^{16}$ cm^{-3}

In the first example:

$$n_c = 5 \times 10^{15} - 10^{15} = 4 \times 10^{15} \text{ cm}^{-3}$$
$$N_T = 6 \times 10^{15} \text{ cm}^{-3}$$

and, hence,

$$\mu = 0.11 \text{ m}^2 \text{ V}^{-1}\text{s}^{-1}$$
$$\sigma = q(4 \times 10^{15}) \times 0.11 \times 10^6$$
$$\sigma = 70.5 \text{ S}$$

In the second example:

$$n_c = 2.5 \times 10^{16} - 2.1 \times 10^{16} = 4.0 \times 10^{15} \text{ cm}^{-3}$$
$$N_T = 4.6 \times 10^{16} \text{ cm}^{-3}$$

and, hence,

$$\mu = 450 \text{ m}^2 \text{ V}^{-1}\text{s}^{-1}$$
$$\sigma = q(4 \times 10^{15}) \times 0.075 \times 10^6$$
$$\sigma = 48 \text{ S}$$

Although the net electron carrier density is the same in both materials, the mobilities differ and therefore conductivities differ.

3.1.3 UNWANTED IMPURITIES

The impurities described above are the simple and technologically important ones. The behavior of other impurity atoms in silicon, for instance, silver, copper, or gold, is much more complex and generally has an undesirable effect on the semiconductor.

Gold, for example, has a valence of one and leaves the bonding deficient of three electrons. Consequently, it will act as a triple acceptor and can be represented on an energy diagram as three discrete levels (see Figure 3.9). Impurity atoms of this kind do not fit into the silicon lattice as well as the simple donor atoms. As a result of the slight distortion that occurs, the energy levels produced generally require more energy to donate or accept electrons and, therefore, are located further from the band edges than the simple dopants, as is shown in Figure 3.9. In general, this type of impurity level is undesirable, and a major problem in semiconductor technology is to remove all such impurities that acci-

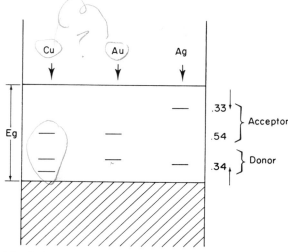

Fig. 3.9 A schematic diagram showing the deep energy levels of three "unwanted" impurities (copper, gold, and silver) in silicon.

dentally find their way into the solid; otherwise their effect might mask those of the desired dopants. Particularly troublesome atoms are copper, carbon, and oxygen. To obtain truly intrinsic behavior in a semiconductor, it is necessary to reduce the concentration of electrically active ions to a value below that of the intrinsic concentration n_i. In silicon, for example, the net doping effect $(N_A - N_D)$ must be less than 10^{10} cm^{-3} (at room temperature)—that is one impurity atom per 10^{12} host atoms! The growth of intrinsic gallium arsenide with its room temperature value of $n_i \cong 9 \times 10^6$ cm^{-3} is beyond the capability of current technology, and the process of compensation we described earlier must be used to obtain "quasi-intrinsic" material.

In a similar situation, defects in the crystal lattice also produce deep-level traps. (A deep-level trap is one removed from the band edges; in fact, near to the center of the forbidden gap.) A missing atom, for example (i.e., a vacancy), results in some of the bonds with the near atoms being unsatisfied, and these local regions will welcome electrons from the valence band. Such a defect would be acceptor-like, thus producing holes (Figure 3.10a). Alternatively, a defect, such as an interstitial, results in valence electrons being redundant from the normal binding process, and they can donate electrons to the conduction band and are donor-like (see Figure 3.10b). In practice, many stable defects that exist at room temperature may result in two or more forbidden levels per defect and, therefore, have a complex influence on the resulting semiconductor. In general, such defects are unwanted, and a great deal of effort is made

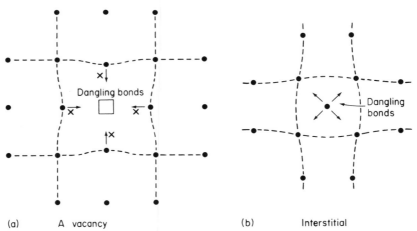

(a)   A vacancy   (b)   Interstitial

Fig. 3.10 Illustrating the lattice relaxation around two simple defects: (a) vacancy; (b) interstitial.

to remove them. There are, however, certain circumstances where this is not true. A prime example is the practical use of radiation damage in gallium arsenide. Crystalline defects, introduced by ion bombardment, produce deep levels in the band gap. These levels can trap and hold free electrons, thus converting n-type gallium arsenide into a compensated semiconductor (such defects also compensate p-type GaAs). Consequently, there are no free carriers, and the material appears to be intrinsic. Because of the low intrinsic conductivity in GaAs, these regions are termed semi-insulating and may be used to isolate devices in a planar technology (see Chapter 7). Another use of deep levels is to provide recombination centers that shorten the lifetime τ of semiconductors. In silicon, both gold impurity levels and radiation damage levels have been used to decrease the lifetime and thus increase the switching speed of silicon thyristors.

3.2 Type Conversion

Once a single crystal is grown, which may be intrinsic, p-type, or n-type, the fabrication of a junction diode or transistor requires that some regions are made p-type and that others are made n-type, with the impurity content very carefully controlled. If a section of crystal is p-type, it can be made n-type by ensuring that the donor density N_D exceeds the acceptor density N_A; this procedure is termed "type conversion" (i.e., we convert p-type to n-type or vice versa). We describe one method of achieving this in Chapter 2, namely to alter the dopant during the growth of a single crystal from the melt. The grown junction method was historically the first junction-type device to be fabricated, but it is rarely used in current planar technology. This method is, however, very valuable for making devices such as lasers.

Planar technology demands greater sophistication in the ways that type conversion is achieved. Both p and n regions with localized three-dimensional complexity are required, and this may be achieved by two techniques: thermal diffusion and ion implantation. Traditionally, silicon devices and integrated circuits have been fabricated by thermal diffusion, and the large capital investment in this technology, made by industry over the last 20 years, will ensure its continued use. A more recent technology is that of ion implantation. Ion implantation is now a highly developed technology and offers many significant advantages over diffusion. For silicon ICs, ion implantation has produced improvements that are significant enough to justify the necessary capital reinvestment needed for a new production line technology. In the case of gallium arsenide, a diffusion technology is effectively nonexistent and ion implantation is the only viable technology for IC fabrication. These considerations will be discussed further later in this chapter.

Consider now the problems of achieving type conversion in some localized region of the semiconductor, as is illustrated in Figure 3.3a. In this example, the base semiconductor is intrinsic silicon and the required dopant is *n*-type (i.e., phosphorus). The object is to take the phosphor from some external source and to place them on sites in the crystal that were occupied by silicon (i.e., to place them on *substitutional sites*). If the dopant impurity does not sit on such a site, its electrical effects may not be the ones that are desired. One example of another undesirable site is the interstitial position shown in Figure 3.10b. Where the dopant becomes attached to a crystalline defect such as a dislocation, or is attached to some undesirable impurity such as carbon or oxygen. Some of these situations are very complex, and the resulting electrical effects are beyond the scope of this book.

A solid like silicon is a very dense concentration of atoms and, thus, work must be done to move the dopant from the surface into the bulk. In other words, energy has to be supplied to the dopant atoms. In atomic diffusion, this source of energy comes from the thermal energy of the diffusion furnace whereas in ion implantation, it arises from the kinetic energy supplied to the ion beam by a particle accelerator. We now discuss these two processes in more detail.

3.3 Atomic Diffusion

To produce atomic diffusion, two conditions are necessary. First, the density of impurity atoms must be nonuniform, with a source of material at or near to the surface and, second, thermal energy is required to enable the atoms to migrate into the solid. To understand the nature of the diffusion and to quantify the process, we must look at its microscopic nature.

In the simplest terms, there are two basic ways in which atoms may diffuse: interstitial diffusion (Figure 3.11a), and substitutional diffusion (Figure 3.11b). (There are, in fact, more ways than this but we will limit our discussion here to these two simple processes.) Consider the first of these two processes. The atom is sitting on the interstitial site, which is a point of minimum potential energy. The surrounding nearest neighbor atoms are forced slightly outwards to accommodate this interstitial (note

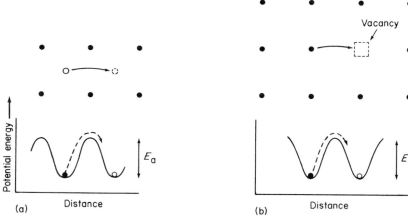

Fig. 3.11 Illustrating the diffusion process and the accompanying potential energy versus distance curves: (a) interstitial; (b) substitutional diffusion.

that the interstitial may be a foreign atom or a self-interstitial, i.e., silicon in silicon). This atom will, like its neighbors, possess kinetic vibrational energy. If at some point this atom has more than the average amount of energy, it may force its way between a neighboring pair of atoms onto one of the next nearest interstitial sites. (The number of such sites in silicon is four and is a geometrical factor that depends on the crystal structure.) To make this jump, the diffusing atom has to do work, and we can show this as a simple oscillating potential energy curve in the lower half of Figure 3.11a. The maximum size of this potential hill is called the activation energy and is denoted by E_a. In many respects, the substitutional diffusion process is similar. Like interstitial diffusion, it can be described by an activation energy E_a. Here, however, a further complication arises. To change from one substitutional site to the next, a vacancy in the lattice must be in the correct neighboring position.

To quantify the diffusion process, let us consider a simple model where the atoms are vibrating with a frequency at the base of a potential barrier of the form shown in Figure 3.11, and where the atoms have a Boltzmann energy distribution. The probability that an atom has an energy in excess of E_a is, therefore, proportional to $\exp(-E_a/kT)$. Since the atom is vibrating with a frequency, it will make two attempts per second to strike the potential hill and pass over it. Furthermore, since, in silicon there are four nearest interstitial sites, then the frequency of jumping is

$$\nu_j = 4\nu \exp\left(\frac{-E_a}{kT}\right) \qquad (3.10)$$

To obtain an order of magnitude estimate of the jump frequency, consider the situation at room temperature (300 K) with a typical experimental value of $E_a \cong 1.0$ eV and $\nu \cong 10^{13}$/s

$$\nu_j = 4 \times 10^{13} \exp\left(\frac{-1.0}{0.026}\right) = 7.92 \times 10^{-3} \text{s}^{-1}$$

that is, an interstitial atom will jump approximately once every 100 s and will rise as the temperature increases to a value typical of that required of diffusion ($\cong 1200°$ C).

For substitutional diffusion, the problem is similar in that if a vacancy is located at a neighboring lattice site, the jump frequency would be $\nu_j = 4\nu \exp(-E_a/kT)$. Here, however, this factor must be multiplied by the probability of finding a vacancy on that particular size. A vacancy is part of a Schottky defect, and the probability of finding this defect is proportional to the density of such defects in the crystal; that is $\exp(-E_s/kT)$, where E_s is the energy required to form such a defect. Thus

$$\nu_j = 4\nu \exp\left[-\frac{(E_a + E_s)}{kT}\right] \tag{3.11}$$

where values of $(E_a + E_s)$ for substitutional impurities in silicon range from 4 to 5 eV and for self-diffusion in silicon $\cong 5.5$ eV. In this case, the jump frequency at room temperature is

$$\nu_j = 4 \times 10^{13} \exp(-4/0.026) = 6.12 \times 10^{54} \text{ s}^{-1}$$

that is, a jump once every 10^{46} years!

The process described above is essentially a random walk. If, however, this mechanism is coupled with the existence of a concentration gradient of a specific atomic species, then a net migration will occur *down* the gradient (i.e., from the higher concentration to the lower). The laws governing this process are called Fick's laws, and a simple derivation is provided in Appendix 4. There are two fundamental equations that may be derived. These are

$$\text{Fick's first law:} \quad \phi = -D\left(\frac{\partial N}{\partial x}\right) \tag{3.12}$$

where ϕ is the net flux of diffusing atoms through a given surface located at x and where the concentration is N and the concentration gradient is $(\partial N/\partial x)$. The D is the diffusion coefficient defined in Appendix 4.

$$\text{Fick's second law:} \quad \left(\frac{\partial N}{\partial t}\right) = D\left(\frac{\partial^2 N}{\partial x^2}\right) \tag{3.13}$$

The second differential equation is very important, since its solution, subject to specific boundary conditions, enables diffusion profiles to be predicted and, hence, enables specific semiconductor device structures to be designed and fabricated. Two types of boundary conditions are generally used, and in the next sections we describe them in detail.

3.3.1 CONSTANT SOURCE DIFFUSION

This condition arises when a semiconductor wafer is subjected to a constant surface flux of diffusing ions. In practice, this occurs when a silicon wafer is placed in a gaseous environment at constant temperature. When this occurs, the solid surface takes up a density of ions appropriate to the solid solubility limit for that species at that given temperature. Some examples of the solid solubility limits are shown in Figure 3.12. If the surface concentration N_0 of the diffusing atoms is held constant throughout the diffusion process, then the resulting profile becomes

$$N(x, t) = N_0 \, \text{erfc}\left(\frac{x}{\sqrt{4Dt}}\right)$$

(erfc is the complementary error function) which, as indicated, is a function of distance x and time as it is shown schematically in Figure 3.13.

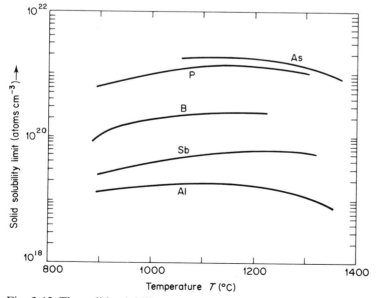

Fig. 3.12 The solid solubility versus temperature data for a range of atoms on silicon [after Colclaser (reference 3, Appendix 6)].

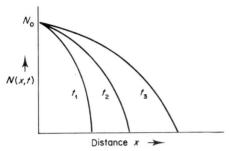

Fig. 3.13 A sequence of constant source diffusion profiles for three times $t_3 > t_2 > t_1$.

Two important points are the following:

a. The temperature dependence on N appears in N_0 (which is a relatively slow dependence) and, more important, is the strong temperature dependence of the diffusion constant D.
b. Diffusion does not give independent control over N_0 and the junction depth. This latter difficulty can be overcome by the use of ion implantation as will be discussed later.

3.3.2 INSTANTANEOUS SOURCE DIFFUSION

In this boundary condition, the total quantity of diffusing atoms Q is held a constant. Consequently, the total area under the diffusion curve (shaded area in Figure 3.14) is a constant. The solution to Fick's second law becomes

$$N(x,t) = \frac{Q}{\sqrt{\pi Dt}} \exp\left(\frac{-x^2}{4Dt}\right)$$

The resulting sequence of profiles obtained at constant temperature but increasing time is shown in Figure 3.14.

In practical device fabrication situations, it is frequently convenient to follow a sequence of diffusions consisting of a source constant short diffusion to place a given quantity of atoms Q into the near surface region, followed by a longer instantaneous source diffusion to push the ions into the solid to produce the actual *pn* junction as is shown in Figure 3.15. This is a very important procedure if a second *pn* junction is required as, for example, in a *pnp* (or *npn*) bipolar transistor. In this way, the surface concentration of the first diffusion can be reduced, thus allowing easy fabrication of the second *pn* junction.

3.3 Atomic Diffusion   55

Fig. 3.14 A sequence of instantaneous source profiles for three times $t_3 > t_2 > t_1$.

Diffusion technology works well with semiconductors such as silicon or germanium. Its application demands only that the crystal can withstand the relatively high temperatures necessary for diffusion without producing any adverse effects. Certain compound semiconductors are examples of solids where the diffusion process proves difficult. In GaAs, for example, long before the diffusion temperature is reached, the arsenic boils away from the surface region of the solid producing a gallium-rich surface unsuitable for device production. A second problem in some compound semiconductors, particularly those whose constituents come from Group II and IV in the periodic table (e.g., ZnS), is autocompensation. Zinc sulphide is partly an ionic solid, thus the zinc and sulphur atoms carry electrostatic charges—the zinc positive and the sulphur negative. To

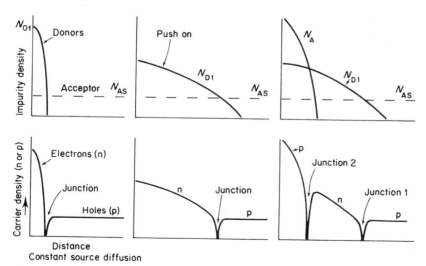

Fig. 3.15 Illustrating the impurity density and carrier density versus distance for the three diffusions needed to fabricate a *pnp* transistor structure.

preserve electrical neutrality, these atoms must exist in the correct proportions. If, for example, a negative sulphur atom is removed from the solid, the localized region of the lattice is endowed with a net positive charge. To ensure that this does not happen, a positively charged zinc atom must be simultaneously rejected (i.e., a zinc vacancy is introduced). This procedure of *automatic compensation* has been proposed to explain the difficulty in producing type conversion in materials like zinc sulphide and cadium sulphide.

3.4 Ion Implantation Doping

In recent years, a technique called ion implantation has been developed as an alternative to diffusion for producing *p*- or *n*-type material. The technique has particular importance when diffusion cannot be used or where the technique offers significant advantages over diffusion. With ion implantation, dopant atoms are injected into the crystal not by the application of thermal energy, but by accelerating the atoms to a high velocity and implanting them into the solid by virtue of their high kinetic energy. The essential feature of this technique is to take a beam of energetic particles emerging from an atom accelerator and to direct them as a collimated beam onto the surface of the solid, as is shown in Figure 3.16*a*.[4] When these energetic ions enter a solid, they are gradually brought to rest by a combination of multiple collisions with the target atoms and electrons in the solid. The collisions with the electrons do not involve any significant transfer of momentum and, hence, do not cause the implanted ion to divert from its straight-line trajectory. It can, in fact, be thought of as a continuous frictional force very similar to a ball bearing moving in a viscous fluid. The collisions between implanted ions and target atoms, on the other hand, will produce deflections in the particle's trajectory, as is shown in Figure 3.16*b*. Furthermore, if the energetic ion imparts more than about 30 eV of energy to the target atom in the collision, the ion can displace it, thus producing a vacancy and an interstitial (i.e., Schottky defect).

These are called radiation-induced defects and must subsequently be removed if the implanted ions are to dope the semiconductor correctly. In practice, this involves heating the semiconductor to elevated temperatures in a clean and nonoxidizing environment. Heating the semicon-

---

[4]With a particle accelerator, ionized atoms are accelerated to some desired velocity by use of an electrostatic potential.

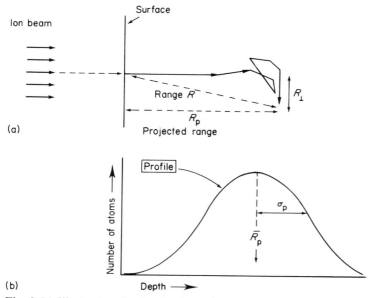

Fig. 3.16 Illustrating the ion implantation process: (a) the trajectory of a typical implanted ion; (b) the range profile resulting from a large number of ions.

ductor to a temperature greater than 600° C does two things: it enables the radiation damage to anneal, and it allows the implanted ions to move onto regular lattice sites of the host crystal, as is shown in Figure 3.17. Here one possibility is shown, that is, an interstitial implanted atom finds its way onto a silicon vacancy. Returning to the actual implanted ions, these will be brought to rest at some distance R (range) inside the solid. The distance R can be divided into its two components, the projected range R_p and the transverse range $R_\perp$. Consequently, even for a monoenergetic ion beam, each of the implanted ions will have a different value of R_p and, hence, the so-called range/profile will have a form similar to that shown in Figure 3.16b. This is, in fact, very close to being a Gaussian curve with a maximum at the mean projected range $\overline{R}_p$ and a standard deviation (called the range straggling) or σ_p. Both are strongly dependent on the mass of the implanted ion, and on its initial velocity, whereas the area under the curve depends on the total number of implanted ions. In Appendix 4, data are provided on the projected range of some selected atoms in silicon and in GaAs as a function of the implanted ion's energy. Also shown is the range straggling σ_p versus energy. These data are of crucial importance to the IC device designer. A second point that must be stressed is that the shape of the implanted profile can be drastically

58 Impurity Doping in Semiconductors

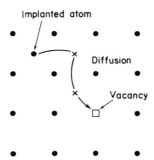

Fig. 3.17 Showing how an implanted ion sitting on an interstitial site moves to a vacancy by annealing and, hence, acts as a dopant atom.

altered if care is not taken to ensure that the incident ion beam is *not* lined up with one of the open channels that exist in single crystal structures.[5] As an example of a typical ion implantation profile, Figure 3.18 shows the resulting three profiles for boron implanted into a silicon crystal at 30 keV, 50 keV, and 150 keV.

A single implantation with a monoenergetic ion beam results in a single Gaussian curve whose projected range $\overline{R}_p$ increases with increasing implant energy. This profile is very limited for device fabrication, since its rigid shape cannot be varied to satisfy some of the complex demands required by the device design engineer. Furthermore, the fall off in doping toward the surface can, itself, be a problem because of the difficulty of making the simple *pn* junction components, as is shown in Figure 3.19. In this instance, the *pnp* structure would result. However, by altering the energy and dose of the implanted ions, the position of a *pn* junction (1) at A in Figure 3.19 can be as carefully controlled as with diffusion. The implantation technique is more powerful than this simple procedure would suggest, since by conducting a series of implantations and by carefully controlling the energy and dose of each implantation, the resulting profile can be tailored at almost any shape. Examples of this concept (Figures 3.20 and 3.21) show two possibilities which can be obtained by summing individual Gaussian curves. One example (Figure 3.20) is a constant doping to some depth where the profile falls rapidly to zero. This could be used to fabricate a simple abrupt *pn* junction. The second example (Figure 3.21) shows a profile that has a large doping close to the surface but that falls off to some constant doping level in the interior of the crystal. Profiles of this kind are very useful in producing variable capacitance diodes (varactors) where the relationship between

---

[5]This phenomenon, called channeling, is discussed in detail in Appendix 6, reference 8.

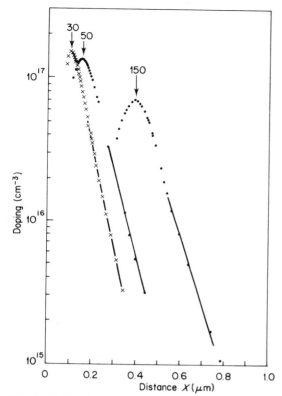

Fig. 3.18 A selection of three boron-implanted profiles in silicon.

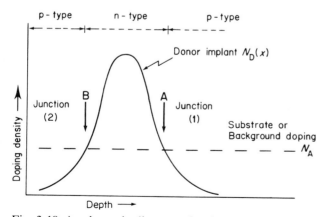

Fig. 3.19 A schematic diagram of a single n-type implant into a p-type substrate, resulting in a *pnp* structure.

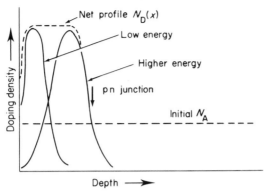

Fig. 3.20 Showing how two implants can be summed to give a flat abrupt *pn* junction.

capacitance and applied voltage depends on the shape of the profile $N(x)$. In practice, it is usual to first define the shape of the desired (ideal) profile and to use a simple iterative procedure, using a computer to determine the minimum number of implantations required to obtain that shape. Thus, the engineer ends up with a specification of the depth $\overline{R}_p$ and the dose of ions required for a sequence of implantations. By using the projected range data in Appendix 4, the depth parameter can be converted into an implant energy. In Figure 3.21 we show a practical result of this process in fabricating a profiled Schottky barrier varactor device in gal-

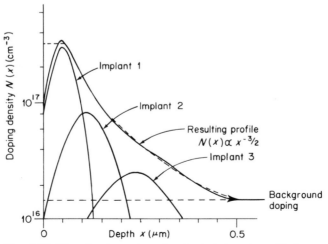

Fig. 3.21 Showing how three implants into *n*-type GaAs can be used to fabricate a profiled GaAs varactor.

3.4 Ion Implantation Doping

lium arsenide. A profile is required starting off at a doping of 3×10^{17} atoms cm^{-3} at the surface and falling off as $N(x) = Bx^{-3/2}$, where B is a constant to a constant level of 1.5×10^{16} atoms cm^{-3}. In this example, the approximation of the desired curve can be obtained by three implantations:

Implantation	Energy (keV)	Dose cm^{-2}
1	57	1.85×10^{12}
2	120	8.97×10^{11}
3	240	4.93×10^{11}

The summation procedure described above is very important because it allows the implantation profile to have independent control over the depth and doping level, a major advantage over diffusion. A second advantage of ion implantation is its ability to obtain good uniformity over large wafer areas and thus good reproducibility from one run to the next. To obtain a good uniformity over the sample area, the ion beam is focused to a spot and is scanned electrostatistically over the sample surface, as is shown in Figure 3.22. The purity and cleanliness is achieved by a combination of an oil-free vacuum system and a mass analysis magnet, which ensures the spectral purity of the implanted beam.

In fabricating complex device structures involving repeated implantations, as is required for integrated circuits, selective area doping is achieved by the use of masks in a manner similar to that used in diffusion. Here, however, it is frequently possible to use conventional photoresist for this purpose. This results in considerable simplification in the fabrication processing, and such simplicity is an excellent way of cutting costs. In this context, implantation offers further advantages over diffusion when one is work with submicron structures. The sideways scatter around

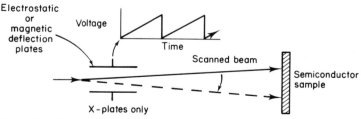

Fig. 3.22 Illustrating how a focused ion beam can be scanned to achieve high uniformity over a crystal surface.

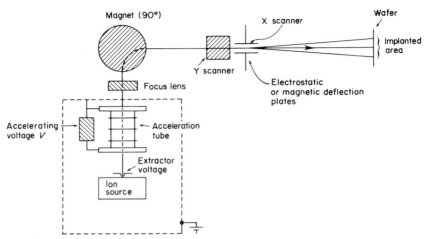

Fig. 3.23 A schematic diagram of a simple ion-implantation facility.

a mask is $\overline{R}_1 \cong \sigma_p$, and is much smaller than the corresponding sideways migration in diffusion. For silicon technology, these advantages are significant enough to have justified the large capital investment required by the electronics industry to install and to operate the ion implantation facilities necessary for commercial exploitation. When, however, we consider compound semiconductors, such as gallium arsenide, ion implantation has another crucial advantage over diffusion. Gallium arsenide is unstable at high temperatures, and surface decomposition occurs when the arsenic is lost. For this reason, there is no diffusion technology available, and the most common alternative for integrated circuit fabrication is ion implantation. These problems are being investigated at the present time with a view to producing monolithic gallium arsenide ICs for very high frequency applications and for very fast logic circuits. In this context, gallium arsenide is preferred to silicon because of its high electron mobility.

To conclude our discussion of ion implantation, we present a brief description of a typical ion implantation facility. Figure 3.23 shows a schematic of an ion implanter.[6] In the ion source, the atoms of the dopant species are ionized and a fine beam is extracted and accelerated through some desired potential (up to a few hundred thousand electron-volts). The accelerated ions are then passed through a magnet that has a field strength so arranged that only those ions with the mass of the desired species can pass through the exit aperture and on to the target chamber.

[6]There are many variations of this particular system depending very much on the company that makes these machines.

3.4 Ion Implantation Doping

The focused beam then passes through an X-Y scanner, which scans the ions in an erasda over the sample wafer. The whole of the ion beam's trajectory occurs in a clean vacuum system, and the target current provides the monitor on the total dose of ions implanted into the semiconductor.

3.4.1 EXAMPLE OF DIFFUSION

A silicon sample has a uniform donor doping level of 3×10^{16} cm^{-3}. The objective is to produce selective p-type doping of the near surface by using boron diffusion. Furthermore, the near surface doping density must be kept below that of the solid-solubility limit of boron, hence, deriving a simple design curve to relate the junction depth to the diffusion time for a drive in diffusion at 1100° C.

Solution: Since it is required to reduce the final surface acceptor concentration to less than solid solubility limit, the final diffusion process must be a drive-in using the instantaneous source diffusion equation:

$$N(x,T) = \frac{Q}{\sqrt{\pi Dt}} \exp\left[\frac{-(x^2)}{4Dt}\right]$$

To do this, we must place Q boron atoms into the near surface region. This can be done in two ways:

Constant source diffusion   or   Ion implantation (near surface implant)

Although ion implantation is the more versatile of these two techniques, in this example we concentrate on diffusion.

3.4.1.1 Predeposition Using Diffusion

In this instance, diffusion is carried out by using the constant source technique; that is, an excess of the dopant is placed at the surface causing the surface to be doped up to the solid solubility limit N_0. Here

$$N(x,t) = N_0 \, \text{erfc}\left(\frac{x}{\sqrt{4Dt}}\right)$$

Note that in some applications where it is required to dope the near surface of the semiconductor p^+ (or n^+), this process can be carried out until the junction reaches the desired depth since, for boron in silicon, the solid solubility limit is reasonably constant above 1000° C. The choice then is to select the temperature where the desired junction may be reached in a reasonable time [not too short (seconds) and not too long (many hours)].

3.4.1.2 Decision: Temperature and Time of Predeposition Diffusion

Some experience is required here. Choosing a value of $T = 1000°$ C places N_0 in the constant region of the boron solubility curve (Figure 3.12). The slightly low temperature ensures that the predeposition does not go too deep. We use

$$T = 1000° \text{ C} = 1273 \text{ K} \quad \text{and} \quad \frac{1}{T} = 0.786 \times 10^{-3} \text{ K}^{-1}$$

and, hence, from Figure A4.2

$$D_1 = 2.7 \times 10^{-14} \text{ cm}^2 \text{ s}^{-1} \quad \text{and} \quad N_0 = 2 \times 10^{20} \text{ cm}^{-3}$$

The total number of atoms Q deposited during predeposition is given by

$$Q = \left[N_0 \left(\frac{4D_1 t}{\pi} \right)^{1/2} \right] \times 10^{-4} \quad \text{atoms cm}^{-2}$$
$$= 3.70 \times 10^{13} \sqrt{t} \quad \text{atoms cm}^{-2}$$

t (s)	1	10	100	10^3	2×10^3
$Q \times 10^{+14}$ (cm^{-2})	0.37	1.17	3.70	11.7	16.5

Times below a few hundred seconds are impractical and we ignored them. Similarly, time above 10,000 s becomes expensive to implement. Thus, we try $t \cong 2000$ s. Hence

$$Q = 16.5 \times 10^{14} \quad \text{atoms cm}^{-2}$$

and

$$\sqrt{4D_1 t} = 1.46 \times 10^{-5} \quad \text{cm}$$

Substituting for N_0 yields $N(x)$ at $t = 2000$ (Figure 3.24). For comparison we note that with ion implantation doping, the total number of atoms predeposited is simply

$Q = \phi$ (the implantation fluence and the peak concentration)

$$N(x) = N_p = \frac{0.4Q}{\sigma_p}$$

The values of σ_p for a particular ion/energy combination can be obtained from tabulated data (see Appendix 5).

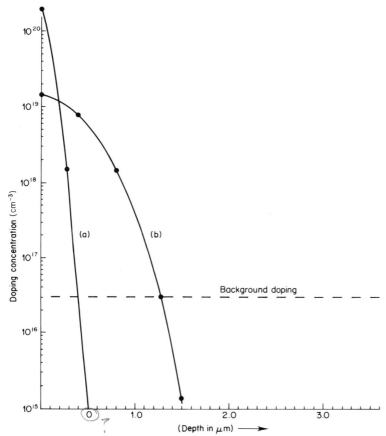

Fig. 3.24 Calculated diffusion profiles of boron in silicon (see text).

3.4.1.3 Drive-in Diffusion

Trial solution: 1100° C for 60 min (3600 s).

There are two conditions to consider. For high temperatures and long times or both, the point where $N(x,t) = N_A$, that is, the junction depth x_j goes deeper while the surface concentration $Q/\sqrt{\pi D t}$ falls. Note that one problem with diffusion is that *there is no independent control* of these two parameters in contrast to ion implantation.

Substituting for D_2 (at 1100° C) = 3×10^{-14} cm² s⁻¹ and

$$t = 3600 \text{ s}$$
$$Q = 16.5 \times 10^{14} \text{ atoms cm}^{-2}$$
$$D_2 t = 1.08 \times 10^{-10}$$
$$(\pi D_2 t)^{1/2} = 1.84 \times 10^{-5}$$

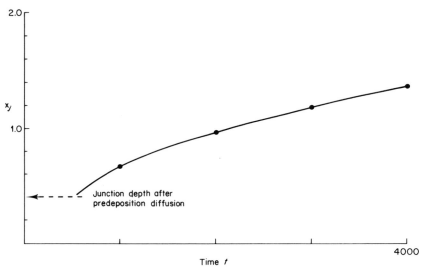

Fig. 3.25 Junction depth versus time for an 1100° C drive in diffusion.

hence

$$N(x) = 8.95 \times 10^{19} \exp\left[-\frac{(x^2)}{4Dt}\right] \text{ cm}^{-3}$$

This is plotted as curve (b) in Figure 3.24. The junction is defined as the point where $N(x_j) = N_D$ and from the graph is approximately 1.29 μm from the surface.

3.4.1.4 Variation of Junction Depth with Time (See Figure 3.25.)

At the actual junction

$$N_0 = N_A(x_j) = \frac{Q}{\sqrt{(Dt)}} \exp\left[\frac{-(x_j^2)}{4Dt}\right]$$

Rearranging we obtain

$$x_j^2 = \ln\left(\frac{Q}{N_D\sqrt{\pi Dt}}\right) 4Dt$$

As a first order approximation[7] we can ignore the dependence of x_j on

---

[7] This is a reasonable assumption since, if t varies from 1 to 10,000, the $\ln(\sqrt{t})$ varies from 1 to 4.6.

$\sqrt{t}$ within the logarithmic term and write

$$x_j^2 = A\,(t)^{1/2}$$

where $A = 2.15 \times 10^{-8}$ (using $t = 3600$ within the log).
As a check, we can substitute $t = 3600$ into the equation and obtain

$$x_j = 1.29\ \mu m \quad \text{(in agreement with figure)}$$

This process can be repeated for different values of temperature, and a whole range of $x_j(t)$ design curves can be obtained.

The other important design consideration is that concerning the surface dopant concentration $N(0)$:

$$N(0) = Q/(\pi D_2 t)^{1/2}$$

Substituting for q and D_2,

$$N(0) = 5.75 \times 10^{25}(t)^{-1/2}$$

Problems

3.1 Determine the value of G for intrinsic silicon at 300 K.

3.2 Determine the value of G for intrinsic germanium at 300 K.

3.3 Determine the value of G for intrinsic gallium arsenide at 300 K.

3.4 What is the band gap energy for an insulator?

3.5 Can SiO_2 be considered an insulator? Why?

3.6 A silicon wafer has an impurity doping density N_D of 10^{16} cm^{-3} at 300 K, what is the resistivity of the wafer?

3.7 A silicon wafer has an impurity doping density N_A of 10^{18} cm^{-3} at 300 K, what is the resistivity of the wafer?

3.8 Determine the mobility μ_n and μ_p for a silicon wafer with a total impurity concentration of 10^{18} cm^{-3}.

3.9 Determine the mobility μ_n and μ_p for a silicon wafer with a total impurity concentration of 10^{16} cm^{-3}.

3.10 Determine the conductivity for the silicon wafer in problem 3.8 with $N_D = 9 \times 10^{17}$ cm^{-3} and $N_A = 1 \times 10^{17}$ cm^{-3}.

3.11 Determine the conductivity for the silicon wafer in problem 3.9 with $N_D = 9 \times 10^{15}$ cm^{-3} and $N_A = 1 \times 10^{15}$ cm^{-3}.

CHAPTER 4

Insulating Films on Semiconductors

Instructional Objectives

This chapter describes the important role of insulating films in semiconductor device manufacturing and device operation. After reading this chapter you should be able to:

a. Summarize the uses of oxide films.
b. Describe how oxides are deposited on semiconductors.
c. Explain how oxides are used in semiconductor device manufacturing.
d. Describe one alternative to an oxide in semiconductor device manufacturing.

Self-Evaluation Questions

Watch for the answers to these questions as you read this chapter. They will help point out the important ideas presented.

a. Why are oxides used as masks in device manufacturing?
b. What is meant by the term "wet oxide"?
c. What is meant by the term "dry oxide"?
d. What happens to impurities in silicon during thermal oxidation?
e. When is Si_3N_4 sometimes used in place of an oxide?

4.1 Introduction

Insulating films play an important role in semiconductor device operation. They may isolate one part of a circuit from another or, in the case of the field effect transistor, prevent direct current flow into the gate while allowing control over the current between the source and the drain. In GaAs, a further important use of insulating films is observed. Arsenic is lost from the surface if the temperature exceeds 500° to 600° C. An insulating film, usually Si_3N_4, is sometimes used to prevent this happening.

In silicon, by far the most common and important film is SiO_2 and, therefore, most of this chapter is devoted to a discussion of it. Some discussion of other types of film is given later in this chapter.

In Chapter 1 we described how an oxide film was used in the formation of our prototype integrated circuit. Its purpose was twofold. First, it was there to protect or passivate the junction between the p and n regions at the point where it came to the surface. The electric field is high in this region and, without the protecting film, the properties of the diode would be vulnerable to changes in humidity and various forms of surface contamination. A second purpose was to prevent the diffusion of the p-type impurity into the silicon, everywhere except in the area selected for the position of the pn junction. In the final process of commercial IC processing a protecting film is usually deposited over the whole circuit with the exception of the bonding areas. This is known as "glassover" and seals and protects the whole integrated circuit.

Thus, we can summarize the uses of oxide films as follows:

a. Passivation of high-field regions on the semiconductor surface.
b. Masking or prevention of diffusion except in selected areas.
c. Masking for selective ion implantation.
d. As the insulating film in the gate region of MOS transistors.
e. As the final circuit protection.

Uses c and d were not discussed in the context of our prototype IC and deserve further mention.

An SiO_2 film can provide a very effective barrier to implanted ions so that if selected areas of the semiconductor need to be ion implanted, an oxide film of sufficient thickness may be used to prevent implantation in those areas where it is not required. Thus, we could replace the diffusion step in the processing of our prototype by ion implantation. The p region would then be formed only in the area required as before.

Formation of the gate region of an MOS transistor represents the most critical use of an SiO_2 film. The MOS transistor depends for its operation on a very high quality SiO_2 layer that separates the gate from the source channel and drain. This use is discussed further in Chapter 7.

4.2 Types of Surface Oxides

In addition to the requirement for oxygen in the formation of SiO_2, silicon is obviously necessary also. Since the films may be formed on high purity silicon wafers, the silicon may, therefore, be supplied by the wafer itself. Alternatively, the silicon may be supplied from some separate source, most usually in a gaseous form. In this circumstance, no consumption of substrate silicon occurs. This difference in the source of the silicon leads to the first major classification of SiO_2 films.

4.2.1 DEPOSITED OXIDES

A deposited oxide is formed when silicon and oxygen are both supplied from external sources. Under the right conditions of temperature, gas flow rates, and pressure, the two elements combine to form SiO_2, which is deposited on the surface. No substrate silicon is consumed in this process.

Referring to the uses of oxide films discussed previously, final circuit protection is an example of the use of a deposited oxide. Clearly it would not be practical to consume silicon for the purpose of circuit protection, since over a large part of the surface the film is not in direct contact with the substrate silicon. In addition, the other type of oxide requires temperatures of around 1000° C for its formation, and this would preclude using it in the final stages of processing.

The technique usually used for the formation of deposited oxides is known as a chemical vapor deposition (CVD). It is a method used for forming many other thin layers, such as single crystal films (a form of expitaxy, see Chapter 2), polycrystalline silicon layers, and films of silicon nitride.

The silicon component of SiO_2 is supplied by silane (SiH_4) and the oxygen by pure O_2. A neutral carrier gas, usually nitrogen, is used. The substrate temperature range is 200° to 500° C, and the total gas pressure is usually close to atmospheric. Films are deposited at a typical rate of 70 nm/min.

Recently, superior film coverage has been found if the gas pressure was reduced to approximately 1 Torr. However, the growth rate is lower

72   Insulating Films on Semiconductors

at approximately 10 nm/min. This technique is called low pressure CVD (LPCVD).

4.2.2 THERMAL OXIDES

If the silicon required to form SiO_2 is supplied by the substrate, then the resulting film is a thermal oxide, so called because it must be grown at high temperature.

The immediate consequences of this technique are:

a. That a separate source of silicon is not needed, thereby simplifying the system.
b. The thickness of the substrate silicon gets less, since silicon is being consumed.

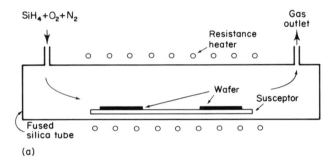

(a)

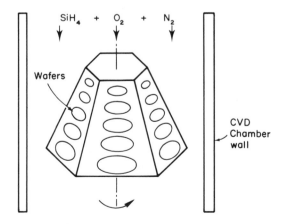

(b)

Fig. 4.1 CVD reactors for SiO_2 deposition: (*a*) horizontal system; (*b*) vertical barrel system for high throughput.

This is illustrated in Figure 4.2. If the film thickness is t, then the substrate thickness is reduced by $0.45t$, whereas the overall thickness, including the film, is increased by $0.55t$.

In the growth of thermal oxides there are two basic means of supplying the necessary oxygen. In a "dry" thermal oxide, the oxygen is supplied as pure O_2 in gaseous form and the reaction is

$$Si + O_2 \rightarrow SiO_2$$

If the oxygen is supplied in the form of steam ("wet" thermal oxide), the reaction is

$$Si + 2H_2O \rightarrow 2H_2$$

The growth rates for wet and dry oxides are quite different and are discussed in the next section.

4.3 Practical Oxidation Systems

Examples of practical oxidation systems are illustrated in Figure 4.1. In this figure, two forms of CVD systems are shown. The substrates are usually mounted on a graphite block or susceptor in a quartz tube, and the susceptor or graphite block is heated by a resistance heater. If uniform thickness films are required, then the substrate temperatures must be uniform over their whole surface. This demands very careful design of the heating system to give a high degree of temperature uniformity over the working length of the tube (the flat zone). An alternative high throughput production system is shown in Figure 4.1.

An example of a thermal oxidation system is illustrated in Figure 4.2. It is capable of growing either wet or dry films by the turn of a control valve. An alternative means of producing high purity water is to pyrolize H_2 and O_2. This produces water of very high purity but has the disadvantage of being more hazardous.

4.4 Mechanisms and Design Rules for Thermal SiO_2 Film

When the oxygen molecule and the silicon atom from the substrate come together to form the SiO_2 molecule, there are, in principle, three ways in which this can occur (see Figure 4.3).

In Figure 4.3a, the oxygen molecule moves through the film and reacts with the silicon at the interface between the film and the silicon. In Figure

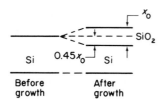

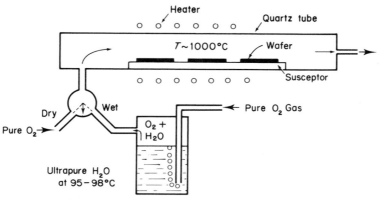

Fig. 4.2 Thermal oxidation system.

4.3b, the silicon moves through the film and reacts with the oxygen at the film surface. Finally, in Figure 4.3c, both oxygen and silicon move into the film, the reaction then occurring somewhere within the film.

Radioactive tracer experiments have indicated that case (a) is really what occurs, the oxidant moving through the oxide film with the formation of the SiO_2 taking place at the interface (Figure 4.3a).

The rate at which a film grows will be governed by two main factors:

a. The rate at which the oxidant moves through the oxide film.
b. The rate at which the chemical reaction to form SiO_2 occurs.

An analysis taking into account both of these factors (see Appendix 5) yields the following relationship between film thickness x_0 and time t:[1]

$$x_0 = \frac{A}{2}\left[\sqrt{\left(1 + \frac{t + t_0}{A^2/4B}\right)} - 1\right] \qquad (4.1)$$

[1]A. S. Grove, *The Physics and Technology of Semiconductor Devices*, Chapter 2, Wiley, New York, 1967.

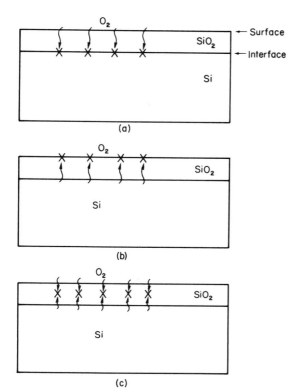

Fig. 4.3 Possible ways in which thermal SiO_2 films may be formed: (a) O_2 moves through film and SiO_2 formed at interface; (b) Si moves through film and SiO_2 formed at surface; (c) O_2 and Si move into film and SiO_2 formed within oxide layer.

where A and B are constants for a given type of oxide at a given temperature. A correction factor t_0, necessary only for dry oxides, is given by

$$t_0 = \frac{x_i^2 + Ax_i}{B} \tag{4.2}$$

where x_i is an initial value of oxide thickness. It is seen from Figure 4.4 that it is nonzero only for a dry oxide.

Equation 4.1 may be rearranged to give the growth time t in terms of the thickness of x_0:

$$t = \frac{A^2}{4B}\left[\left(\frac{2x_0}{A} + 1\right)^2 - 1\right] - t_0 \tag{4.3}$$

The constants A and B are extremely sensitive functions of temperature and are given by

$$A = K_1 \exp\left(+\frac{E_1}{kT}\right) \tag{4.4}$$

where K_1 and E_1 depend only on the type of oxidation employed (i.e., wet or dry), k is Boltzmann's constant, and T is the absolute temperature.

$$B = K_2 \exp\left(-\frac{E_2}{kT}\right) \tag{4.5}$$

E_1 is the activation energy of the diffusion process. $E_1 + E_2$ is the activation energy of the reaction process.

The parameters K_1, K_2, E_1, E_2, and t_0 are given in Figure 4.4 for both wet and dry oxidation. The numerical values of parameters for oxide thickness calculations for $\langle 111 \rangle$ and $\langle 100 \rangle$ silicon are presented in Table 4.1. Using these values in conjunction with Eq. 4.1 or 4.4 allows the calculation of any oxide thickness for a given temperature and time or, alternatively, the prediction of the time for a given oxide thickness.

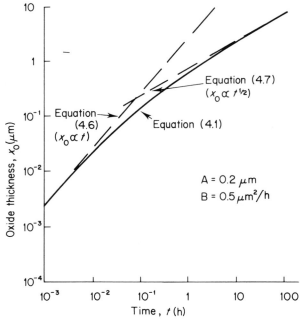

Fig. 4.4 Approximate and accurate forms for the dependence of oxide thickness on time.

TABLE 4.1 Numerical Values of Parameters for Oxide Thickness Calculations

Parameter	⟨111⟩ Silicon			⟨100⟩ Silicon		
	Wet	Pyrolytic	Dry	Wet	Pyrolytic	Dry
K_1 (μm)	2.39×10^{-6}	2.37×10^{-6}	1.24×10^{-4}	4.02×10^{-6}	3.98×10^{-6}	2.08×10^{-4}
K_2 (μm^2/hr)	214	386	772	214	386	772
E_1 (eV)	1.29	1.27	0.77	1.29	1.27	0.77
E_2 (eV)	0.71	0.78	1.23	0.71	0.78	1.23
x_1 (μm)	0	0	0.02	0	0	0.02

$k = 8.62 \times 10^{-5}$ eV/K.

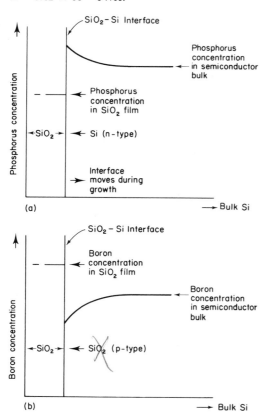

Fig. 4.5 Impurity distribution at oxide interface: (*a*) phosphorus impurity; (*b*) boron impurity.

Practical examples are given below to illustrate the calculations involved. However, it is instructive to consider two limiting cases of Eq. 4.1.

For film thicknesses that are very small (i.e., during the early stages of growth) the oxidant moves through the film very readily, and the limiting factor will be the rate at which SiO_2 molecules are formed at the interface.

The growth in this instance is said to be "reaction rate limited." If $t + t_0 \gg A^2/4B$, then Eq. 4.1 reduces to

$$x_0 = \frac{B}{A}(t + t_0) \tag{4.6}$$

The constant B/A is known as the linear rate constant.

As the film thickness increases, it becomes progressively more difficult for the oxidant to move through the film, and the rate at which it can do so becomes the limiting factor. The growth rate is then diffusion rate limited. For $t + t_0 \gg A^2/4B$, Eq. 4.1 reduces to

$$x_0 = (Bt)^{1/2} \tag{4.7}$$

The parameter B is known as the parabolic rate constant.

The film thickness-time relationship is shown schematically in Figure 4.5, where the x_0 and t are plotted on a logarithmic scale. The two regions corresponding to Eqs. 4.6 and 4.7 are clearly indicated.

4.5 Additional Effects in the Oxidation Process

4.5.1 ENHANCED GROWTH AT HIGH DOPING LEVELS

When thermal oxides are grown on heavily doped silicon, the growth rates are increased above those discussed for more moderately doped silicon (see Section 4.4). We next consider separately the effects for the two major n- and p-type dopants, phosphorus and boron.

In phosphorus doped silicon, the growth-rate enhancement is not significant below 10^{20} cm^{-3} and above temperatures of 1000° C.

Enhancement is greater for wet oxides than for dry oxides, and films can be approximately four times as thick at 900° C, based on the calculations of Section 4.4.

4.5.2 EFFECT OF HCl

As we discussed in more detail in Section 4.6, the quality of an oxide can be improved considerably if the oxidation is carried out with the addition of a few percent of HCl when growing dry oxides. However the growth rates are altered if HCl is used, and the film thickness will be approximately one and one-half times greater if 10 percent HCl is used.

4.5.3 IMPURITY REDISTRIBUTION

If silicon is consumed in the formation of SiO_2 during thermal oxidation, then the question arises as to what happens to any n- or p-type impurities that are present in the silicon. Are they incorporated into the SiO_2 film along with the Si atoms, or are they pushed ahead of the moving Si-SiO_2 interface, thereby building up in the silicon?

The behavior turns out to be different for different types of impurities, and Figure 4.6 illustrates the effects for the two major dopants, phosphorus and boron.

In the case of phosphorus, a smaller amount is incorporated in the film than exists in the semiconductor with the result that the remainder builds up in the silicon close to the interface.

With boron it is different. A greater amount is incorporated in the film than exists in the semiconductor bulk, so that the silicon close to the interface becomes depleted of the impurity.

Clearly, this effect must be taken into account in the design of any semiconductor device in which the impurity concentration in the semiconductor near the interface is important (an MOS transistor is a good example).

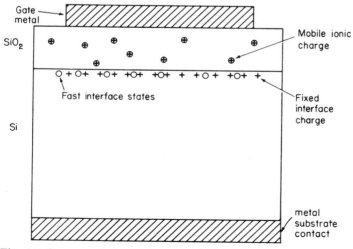

Fig. 4.6 Sources of charge in a practical MOS system.

4.5.4 HIGH PRESSURE OXIDATION

When the oxide growth is diffusion-rate limited, as is the case for thicker oxides, the rate at which the oxidant diffuses through the film can be increased simply by increasing the concentration gradient in the oxide film. This may be accomplished by increasing the impurity concentration at the surface, N_S (see Appendix 5) which, in turn, may be achieved simply by increasing the partial pressure of the oxidant during growth. Thus, an increase in growth rate can be achieved by carrying out the oxidation with the gas pressures at several atmospheres.

4.6 Assessment of Film Quality

The most critical application of a thermal oxide film is as the gate oxide in an MOS transistor. Thus, the assessment of its quality is usually made in this context. Figure 4.6 illustrates the three important factors that can degrade the properties of an MOS transistor when they are present in its gate oxide.

First, there is the fixed charge that exists at the interface of any thermally grown layer. It is independent of any potential applied to the transistor and is usually positive. Its magnitude is usually higher in a wet oxide than a dry, especially above 900° C growth temperature. However, it can be reduced by up to an order of magnitude by a subsequent anneal in nitrogen or argon. The interface state density may also be reduced if a small percentage of HCl is incorporated in the gas stream during dry oxide growth. However, this can affect the growth rates and should be taken into account (see Section 4.5).

An estimate of the magnitude of the interface state density can be made by measuring the capacitance of an MOS capacitor as a function of voltage. The presence of interface states causes a shift of the C-V curve along the voltage axis and, by measuring this shift, the estimate can be made.

The second type of defect that can occur in the oxide film is the presence of mobile ions, which will move when a gate potential is applied over a period and thus change the characteristics of the transistor. These are illustrated in Figure 4.6. They are usually present because of sodium contamination and are not likely to occur in significant amounts. The presence and magnitude of these mobile charges may be assessed by making capacitance-voltage measurements before and after a high temperature stress procedure in which the gate is held positive or negative while the device is held at a temperature of 200° to 250° C. This causes the mobile ions to move to the side of the oxide that is at the negative potential, the high temperature speeding up this process. If there are mobile ions present in the oxide, this will show up as a shift along the voltage axis of the C-V curve between the two measurements. The magnitude of the ionic charge can be inferred directly from the magnitude of the voltage shift.

The third type of defect is known as the fast interface state. This occurs near the oxide-semiconductor interface and is able to trap and release charge depending on the bias conditions. The capacitance-voltage relationship in this instance is shifted by varying amounts on the voltage axis. Thus, the C-V curve is distorted and, by analyzing the degree of distortion, the density of the fast interface states may be calculated.

4.7 Choice of Film Type

The different types of film discussed have contrasting properties and it is, therefore, to be expected that different types are used depending on the application to which they are put.

It has already been argued that a deposited CVD film is more appropriate for final circuit protection. The rest of this section is devoted to a discussion of the relative applications of wet and dry thermal oxides.

The initial oxide in an MOS process is usually fairly thick (~ 1 μm) and is usually grown as a wet oxide. The reasons for this are the following:

a. Quality requirements are not stringent.
b. High growth rates are desirable to form film in a reasonable time period.

In contrast, the gate oxide in an MOS process does have stringent quality requirements, and the thickness required is typically 0.1 μm. The lower growth rates for a dry oxide still result in reasonable growth times, so it is usual for gate oxides to be grown using dry O_2.

4.8 Nitride Films

When one considers the use of an SiO_2 film as a mask to prevent the diffusion of impurities into unwanted areas, it must be kept in mind that impurities will still diffuse to some extent into the SiO_2 film and what makes its use as a mask possible is the relative diffusion rates of any particular impurity in Si and SiO_2. The two major dopants, phosphorus and boron, diffuse relatively slowly in SiO_2 compared with silicon and, hence, SiO_2 can provide effective masking. Of course, the film must be of adequate thickness as discussed in Section 4.5. However, there are dopants that diffuse at a relatively high rate in SiO_2, so that the latter may not be used as a mask. Examples of these dopants are Ga, Al, Zn, Na, and O_2.

It is found that silicon nitride (Si_3N_4) is an effective mask for these impurities.

Silicon nitride is usually deposited by the CVD technique so that no substrate silicon is consumed in its formation. The reaction between silane and ammonia is utilized (see Section 2.4, Chapter 2) to form the nitride layer. Substrate temperatures are usually between 800° and 1000° C, with a typical growth rate at 800° C being 3 nm/min.

Nitride films give rise to a high density of interface states; therefore, it is usual to first grow a thin SiO_2 film and then deposit the Si_3N_4. This system is used in a special type of MOS transistor known as the MNOS transistor.

The gate region consists of a thin grown layer of SiO_2 covered with a relatively thick layer of Si_3N_4. Defects that exist at the interface between the SiO_2 and Si_3N_4 can be charged or discharged by the application of fairly high potentials. The charge state can remain indefinitely under normal working voltages.

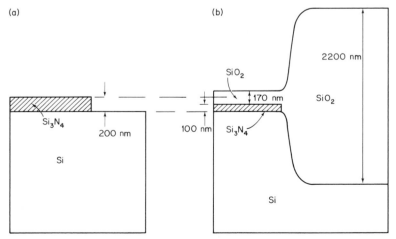

Fig. 4.7 SiO_2 growth over silicon nitride film (idealized profile): (*a*) before; (*b*) after.

It is sometimes necessary to grow an SiO_2 layer over a region partly covered by a nitride film (see Figure 4.7). In this circumstance, the SiO_2 growth over the nitride film takes its silicon from the latter, thereby converting it to SiO_2. For every micrometer of Si_3N_4 converted, 1.7 μm of SiO_2 is produced. This conversion rate is much slower than the SiO_2 growth rate on the silicon; hence, for example, during the conversion of 0.1 μm of Si_3N_4 in steam at 1100° C (producing 0.1 μm of SiO_2 over the nitride), more than 2 μm of SiO_2 is grown over the silicon.

Example 4.1

A dry thermal oxide 0.2 μm (0.2 × 10^{-4} cm) thick is required to be grown at 1100° C on ⟨111⟩ orientation silicon. Calculate the time required.

Solution

T (in degrees Kelvin) = 1100 + 273 = 1373 K. Using the value of k in Table 4.1, kT = 0.119 eV. Using Eqs. 4.2, 4.3, 4.4, and 4.5

$$A = 1.24 \times 10^{-4} \exp(0.77/0.119) = 0.08 \ \mu m$$
$$B = 772 \exp(-1.23/0.119) = 0.025 \ m^2/hr$$
$$t_0 = [0.02^2 + A(0.02)]/B = 0.082 \ hr$$

Inserting these values into Eq. 4.3

$$t = 2 \ hr \ 10.4 \ min$$

If the above example is repeated for a temperature of 800° C, then 130 hr oxidation time is necessary. This is obviously impractical, especially for a commercial process, and is one of the main reasons that typical oxidation temperatures are usually near 1000° C.

Example 4.2

An oxide is grown for 130 min at 1100° C on ⟨100⟩ silicon by passing oxygen through a 95° C water bath. Calculate the resulting oxide thickness.

Solution

Again, using Table 4.1 and Eqs. 4.1, 4.2, 4.4 and 4.5

$$A = 4.02 \times 10^{-6} \exp(1.29/0.119) = 0.205 \ \mu m$$
$$B = 214 \exp(-0.71/0.119) = 0.549 \ \mu m^2/hr$$
$$t_0 = 0 \ \text{for wet oxides}$$

Substituting in Eq. 4.1

$$x_0 = 1 \ \mu m$$

A comparison of the results of the above examples shows that wet oxides grow much faster than dry oxides. An examination of the activation energies in the diffusion limited region for wet and dry oxides shows that E_2 is nearly double that for a dry oxide. However, the pre-exponential factor K_2 is greater for a dry oxide, implying a higher growth rate. Thus, the lower growth rate for the dry oxide occurs because of the higher activation energy.

Problems

4.1 Why is nitrogen used as a neutral carrier gas with silane and oxygen to produce silicon dioxide?

4.2 If an oxide film is deposited at a rate of 70 nm/min, how long would it take to deposit 15,000 Å?

4.3 If a silicon wafer has a thickness of 0.4572 mm before oxidation, determine: (a) the thickness of the silicon after 1000 Å of SiO_2 is grown by thermal oxidation; (b) the total thickness of the silicon and the oxide after the 1000 Å oxide is grown.

4.4 How many millimeters of silicon is consumed to grow 15,000 Å of SiO_2?

4.5 Calculate the time required to grow a thermal oxide 0.01 μm (1000 Å) thick on $\langle 111 \rangle$ silicon at 1075° C.

4.6 Calculate the time required to grow a dry oxide 0.01 μm (1000 Å) thick on $\langle 100 \rangle$ silicon at 1075° C.

4.7 Calculate the time required to grow a wet oxide 0.01 μm (1000 Å) thick on $\langle 111 \rangle$ silicon at 1075° C.

4.8 Explain the term "pyrolytic" oxide.

4.9 An oxide is grown for 65 min at 1075° C on $\langle 111 \rangle$ silicon by passing oxygen through a 95° C water bath. Calculate the resulting oxide thickness.

4.10 A dry thermal oxide is grown with oxygen and 10 percent HCl. How much more oxide will be grown over that of an oxygen only system?

CHAPTER 5
Photolithography

Instructional Objectives

This chapter introduces the fabrication steps for a simple integrated circuit. After reading this chapter you will be able to:

a. Describe a negative photoresist process sequence.
b. Describe the difference between negative and positive photoresist.
c. Describe photolithographic glass slides.

Self-Evaluation Questions

Watch for the answers to these questions as you read this chapter. They will help point out the important ideas presented.

a. When is photolithography used?
b. When are positive and negative photoresists used?
c. What problems occur when photoresist films are overexposed?
d. Why must the photoresist film have uniform thickness?

5.1 Introduction

In Chapter 1, we examine the need for two pattern-forming stages or masking levels in the fabrication of a simple prototype integrated circuit. In a more complex circuit, more masking levels are usually necessary.

86   Photolithography

In each instance, to carry out batch processing, the pattern for each individual circuit must be repeated over an array that is large enough to cover the entire wafer. Several requirements that must be met by the technique used for forming the patterns can be identified:

a. It must be suitable for forming patterns on different types of surface film. In the previously presented example the first pattern was formed in SiO_2 whereas the second was formed in aluminum.
b. It is necessary to be able to "align" each pattern level accurately with the preceding one.
c. The dimensional accuracy of each individual pattern must be sufficient to ensure proper alignment between the levels.
d. The chip size or repeat distance must be accurately maintained between different levels, since any error would accumulate across the array and lead to misregistration on some circuits within the array.

The technique generally used in the semiconductor industry for all pattern formation is known as lithography. The most common form of lithography uses ultraviolet (UV) light and is called photolithography. Other types are electron beam and X-ray lithography. To illustrate how lithography is applied, the example of Chapter 1 will be used to show how the first-level pattern in the silicon dioxide (the opening for the p^+ diffusion) would be produced. It will become apparent how the above requirements are satisfied.

The process has similarities with conventional photography in that there is a light-sensitive film (photoresist), or emulsion, a controlled exposure to an image, and a development. A safe-light environment for carrying out the processing is also necessary to prevent overexposure and loss of image. Figure 5.1 shows the various stages in the process.

(i)   The silicon wafer is first oxidized (see Chapter 4).
(ii) The wafer is coated with photoresist.
(iii) A glass "mask" containing the pattern for the appropriate level is brought into contact with the wafer.
(iv) The wafer with photoresist is then exposed for the correct exposure time to a collimated beam of UV light, then the mask is removed.
(v)   The resist film on the wafer is then developed. At this stage, the mask pattern has been reproduced in the photoresist, but as a negative.
(vi) An etch must be used that will remove SiO_2 but will not significantly attack the photoresist film or the underlying silicon. When the SiO_2 in the uncovered areas have been removed, the final objective of

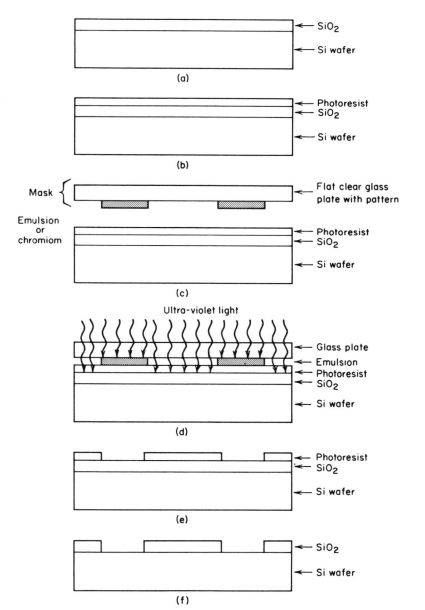

Fig. 5.1 Process sequence in negative photolithography: (a) growth of SiO_2 film; (b) coating with photoresist; (c) mask placed in proximity; (d) mask aligned and brought into contact; (e) photoresist developed; (f) oxide etch and photoresist removed.

producing a pattern of "holes" in the SiO_2 for the subsequent p-diffusion has been achieved.

Now, we describe the level 2 pattern in aluminum. The process is similar to that described above but would have the following modifications:

(i) The underlying film would be aluminum rather than SiO_2.
(ii) The SiO_2 selective etch used in Step (vi) would be replaced by an aluminum-selective etch.
(iii) A different mask would be used with the second level pattern on it.
(iv) The second pattern would have to be aligned with the first so that the aluminum contact lies in the correct position over the p-region window.

The latter requirement is extremely important and is discussed later in the section on mask alignment.

5.2 Photoresist Types

Photoresists are supplied in the form of a liquid and consist of some ultraviolet sensitive substances in a solvent base. If we compare the mask and substrate in Figure 5.1 (iii) and (v), we can observe that the pattern produced in the photoresist is the inverse or negative of that on the mask. For this reason, those resists that are not removed on exposure to ultraviolet light and development are known as negative resists. Other types of resist, which are positive working, are removed on exposure and development. A comparison between positive and negative resists is made in Figure 5.2, where the same final pattern in SiO_2 is produced by the alternative techniques.

Negative resists consist of a base of synthetic rubber compounds containing a few percent of a light-sensitive agent that facilitates the formation of cross linkages between base modules. This cross linking inhibits its dissolution so that exposed areas will remain while unexposed areas will be removed in an appropriate solvent.

Positive resists have a quite different chemistry, a light-sensitive compound, in this case, inhibiting dissolution unless the action of light has previously broken up the compound. In this circumstance, exposed areas are removed and unexposed areas remain.

If photoresist films are underexposed, there is a tendency for the pattern formation to be incomplete and, in the extreme case, to cause a

5.2 Photoresist Types

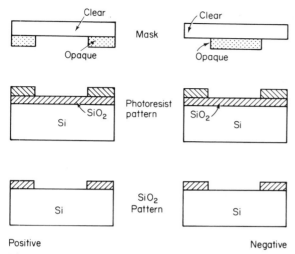

Fig. 5.2 Positive and negative resists.

total loss of pattern. With positive resist, the film remains intact whereas, for a negative resist, all is removed.

If the film is overexposed, then windows opened in positive photoresist are slightly larger than the mask dimension. This is due to scattered light penetrating under the mask edges and exposing a small region of film not directly irradiated by the light source. Subsequent etching of the

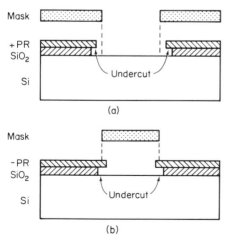

Fig. 5.3 Changes in window sizes with positive and negative resists: (*a*) positive resist; (*b*) negative resist.

underlying SiO_2 film accentuates this enlargement, as is shown in Figure 5.3a.

With negative resists, the windows tend to be smaller than the mask dimension, as is illustrated in Figure 5.3b, but this is partly compensated for during the etching of the SiO_2.

5.3 Film Thickness

One of the important functions of the photoresist film is to protect underlying films of varying types (i.e., SiO_2, aluminum, polysilicon, and silicon nitride). Thus, it must be thick enough to provide an effective barrier to the various etches to be used. Also, in the case of selective ion implantation, it has to prevent ions from reaching the underlying surface regions.

This sets a lower limit on film thickness. An upper limit is set by the need for good definition in the pattern, as is shown in Figure 5.4. It is unlikely that sidewalls will be precisely vertical following development, and any uncertainty will clearly have a greater effect as a percentage of the nominal window dimension, depending on the thickness of the film. Thus, the technique for coating wafer surfaces with photoresist must be capable of providing very uniform films with a high degree of control over the thickness.

The way in which thin, uniform films are obtained is illustrated in Figure 5.5. Drops of liquid photoresist are introduced onto the slice held by a vacuum check, which can be set to rotate accurately at some pre-

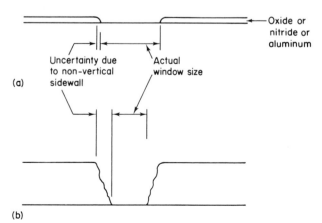

Fig. 5.4 Effect of nonvertical sidewall definition for (a) thin; (b) thick resist films.

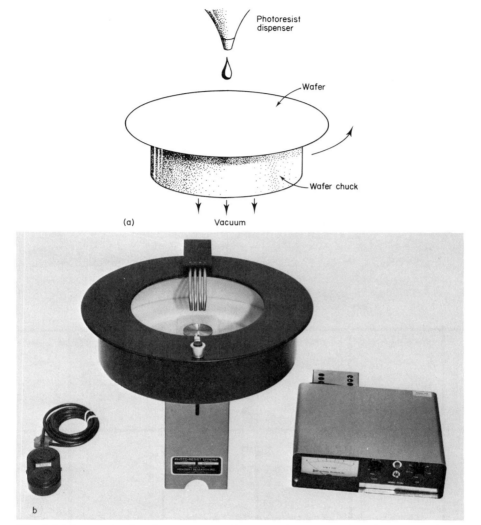

Fig. 5.5 Photoresist film formation: (*a*) Coating with photoresist; (*b*) commercial photoresist spinner (courtesy Headway).

determined speed. The slice is then spun, centrifugal forces spreading the liquid outward, with any excess being thrown clear of the wafer. After a certain time, the film thickness will stabilize at a value dependent on the rotational speed. The slice is then removed and baked to drive off the solvent and to form a solid film that is thin and uniform over most of the slice. Some increase in thickness will inevitably occur at the edge of the slice as a result of surface tension in the liquid. Thus, it may be

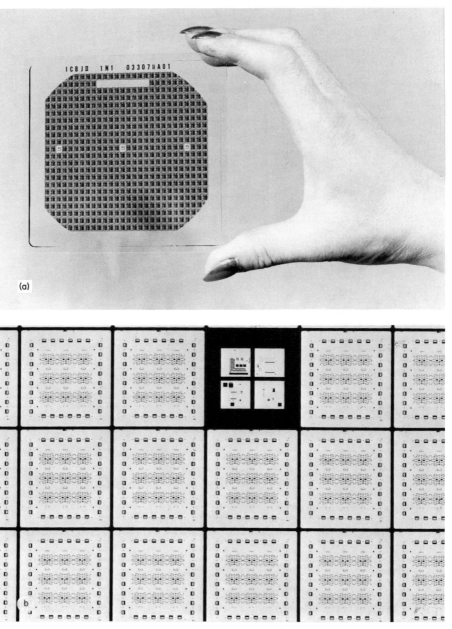

Fig. 5.6 Mask plate with enlarged view of one circuit: (*a*) mask photography (courtesy Siliconix Ltd, Swansea); (*b*) enlarged pattern of above mask.

necessary to discard later some of the circuits very close to the edge. Figure 5.5 also illustrates a commercial 1-head spinner that has accurate speed control, with very high acceleration and deceleration at the start and finish of the spin cycle.

5.4 Masks and Mask Making

The prerequisite in the photolithographic process is a mask set, each containing a uniformly repeated pattern appropriate to the particular processing level for which it is intended. A photograph of a typical mask is shown in Figure 5.6 with an enlarged view of one of the patterns also illustrated. The mask consists of an ultra-flat square glass plate, typically 1 mm thick with one face covered with the pattern. It is, of course, this

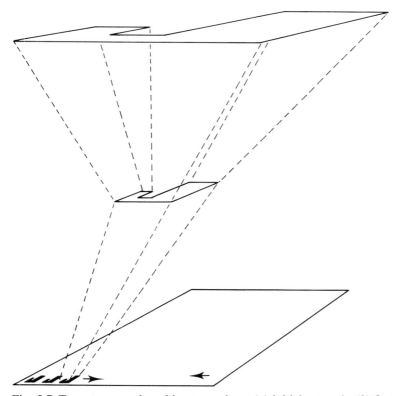

Fig. 5.7 Two-stage mask making procedure: (*a*) initial artwork; (*b*) first reduction ($\sim \times$ 10 to 40); (*c*) step and repeat ($\sim \times$ 10); (*d*) mask.

face that always must be brought in contact with the coated slice during the exposure.

The opaque regions are usually chromium, and the patterns are formed by a photolithographic technique similar to the one discussed previously. However, an additional complication present in mask making is that the pattern images have to be repeated in precise steps.

The way in which masks are produced is illustrated in Figure 5.7. The integrated circuit designer decides on the number of masks needed for the complete process and produces carefully dimensioned sketches of each mask. When the complexity of the integrated circuit is not very high, the pattern is cut, say at a few hundred times actual size, in double-skinned mylar sheet. Mylar is a special plastic film which is dimensionally very stable. This mylar is called Rubylith.

One skin is clear and the other is usually red and, therefore, photographically opaque. The pattern is cut in the red film and peeled away where clear areas are desired. The cutting is usually carried out on a precision table where the x and y positions of the cutter can be very accurately controlled. Only one pattern of the array is cut, and the re-

Fig. 5.8 Kasper mask aligner.

sulting sheet at this stage is called the initial artwork. The pattern is reduced by a factor of 10 or more, using a reduction camera, and the result is an intermediate photographic plate containing one pattern 10 times larger than that finally required.

The final reduction is combined with a precise stepping of the pattern in a piece of apparatus known as a step and repeat camera. This is illustrated schematically in Figure 5.7.

An alternative mask-making technique, which is increasingly used particularly for high density circuits, is for the designer to "digitize" the required patterns on the screen of a special purpose computer. The patterns are then stored on tapes, which are used to drive a mask-making machine directly. An obvious advantage of this approach is that standard designs may be stored and, if necessary, modified at will without having to write off any earlier effort and start fresh each time. Also, it enables the designer to check his or her design with greater accuracy, since the facility to view any portion of the overall circuit is readily available on such machines (see Figure 5.8).

Problems

5.1 Why does photolithography use ultraviolet light?

5.2 What is meant by the term "safe-light environment" when working with photoresist?

5.3 Describe what might happen if photoresist films are underdeveloped.

5.4 What materials are usually used to make photolithographic masks?

5.5 What can be used as an alternative to a glass mask when the complexity level of the circuit is small?

CHAPTER 6
Metallization, Interconnections, and Packaging

Instructional Objectives

This chapter presents the techniques used to interconnect and package semiconductor devices. After reading this chapter you will be able to:

a. Describe how metal interconnections are made in integrated circuits.
b. Describe how a semiconductor device is protected from chemical contamination.
c. Describe how semiconductor devices are mounted in packages.
d. Describe how wires are connected to integrated circuits.

Self-Evaluation Questions

Watch for the answers to these questions as you read the chapter. They will help point out the important ideas presented.

a. Explain how the source material can be heated in a vacuum deposition system.
b. What advantage does electron beam evaporation have over resistance heating evaporation?
c. Why is electrolytic deposition inferior to vacuum deposition?
d. Why is SiO_2 deposited over an entire microcircuit?
e. Describe the difference between thermocompression and ultrasonic compression.

6.1 Introduction

The semiconductor device structure or integrated circuit has no practical use until it is connected to the outside world. In practice, this means that electrical terminals of high conductivity must be attached to the device so that voltages can be placed across the device and current can be drawn from it. Usually, this necessitates that a low resistance ohmic contact be made to the device terminals (an ohmic contact is an ideal situation where the metal to semiconductor interface has no resistance). The majority of all devices fabricated are with planar technology, where all the component devices, transistors, diode resistors, capacitors, and inductors are embedded in or near the surface region of the semiconductor. Consequently, all of the metal contacts and interconnections must be deposited on the one surface.

6.2 Thin Film Deposition

With planar technology, the fabrication of contacts and interconnections by thin film deposition is very attractive. Vacuum deposition combined with photolithography can provide this technology. There are two basic procedures for depositing thin films: vacuum evaporation, and sputtering.

6.2.1 VACUUM EVAPORATION

A schematic diagram of a vacuum evaporator is shown in Figure 6.1. The material to be evaporated is placed in a small heater. The temperature of the heater is then raised until the source material vaporizes and atoms will be evaporated in all directions. These atoms will condense on the inside surface of the bell jar and the substrate. By placing an evaporation mask in front of the substrate, the geometry of the evaporated film can be controlled. There are many variations of this basic technique to suit the different materials; for example, the source material can be heated by resistance heating, or alternatively by electron bombardment heating.

A large variety of the evaporation boats used in resistance heating are available commercially. Some examples are shown in Figure 6.2. The most important criterion is that the boat metal must have a melting point that is much higher than that of the metal being evaporated. Usual choices are tungsten or molybdenum (with melting temperatures of 3370° C and 2620° C, respectively). Boat geometries can range from the simple coiled wire, or indented ribbon, to complex structures resembling small ovens.

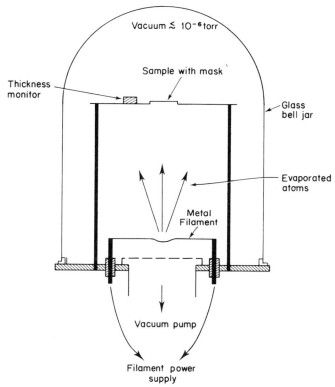

Fig. 6.1 A schematic diagram of a vacuum deposition unit.

One very useful boat available commercially consists of a cone-shaped coil of wire, covered with alumina (Figure 6.2). This works particularly well with gold, silver, and the alloy of gold and germanium.

Electron-beam evaporation uses a focussed electron beam of high intensity to vaporize the source material. A schematic diagram showing the electron-beam evaporation procedure is shown in Figure 6.3. This technique is extremely powerful, partly because of its potential purity and, equally important, because of the much wider range of materials that can be evaporated in this way. This is particularly important with the high melting point metals such as platinum and tungsten where resistance heating cannot be used. Another important feature of the electron-beam technique is that the vaporization occurs only at a small spot in the source material. The supporting hearth is not directly heated and, hence, does not emit unwanted contaminant atoms as occurs in the conventional resistance heated boats. Any foreign atoms absorbed on such boats will be evaporated along with the desired metal.

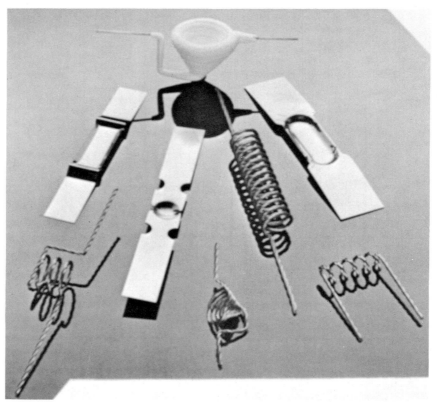

Fig. 6.2 Illustrating a range of evaporative filaments available commercially (permission of Nordico Ltd).

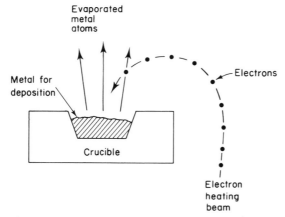

Fig. 6.3 A schematic diagram of an electron beam evaporative unit.

6.2.2 SPUTTERING

The essential features of the sputtering technique are illustrated in Figure 6.4a. The bell jar is evacuated and an inert gas, such as argon, is bled through the needle valve to maintain a background pressure of $\cong 10^{-2}$ Torr. The cathode is made of the material that is to be deposited onto the substrate. With the application of a high voltage (2–6 kV) between the electrodes, the inert gas is ionized and the positive ions are accelerated to the cathode. On striking the cathode, these ions will collide with the cathode atoms, giving them sufficient energy to be ejected. These sputtered atoms will travel through space, eventually coating the anode and, of course, the substrate.

The use of masking techniques can again be used to localize the deposition of desired areas. The simple sputtering system described above is a diode sputtering system. Practical coating units are, in general, more complex in detail (Figure 6.4a) but work on the same basic physical mechanism. By using the more complex sputtering systems, it is possible

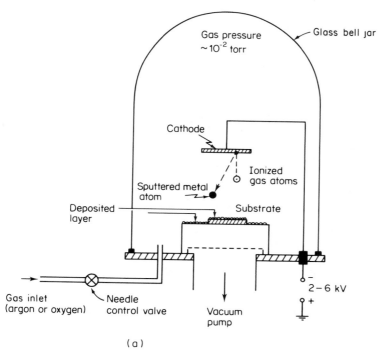

Fig. 6.4 A diode sputtering unit: (a) schematic diagram.

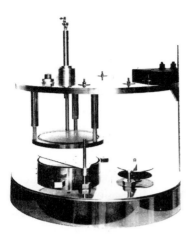

Fig. 6.4 (*b*) a real system (permission of Nordico Ltd).

Fig. 6.4 (*c*) Sloan SL1800 thin film deposition system with electron beam sources.

to sputter and deposit a range of metals as well as dielectric layers, such as silicon dioxide.

6.3 Plating

In comparison with the vacuum deposition techniques, plating is a much simpler and less expensive technology. Plating is essentially a process of the electrolytic deposition of metal from solution and there are two basic variants: electrolytic and electroless. With electrolytic deposition, the surface on which the metal is to be deposited is used as the cathode, and deposition occurs when the metal ions travel under the influence of the electric field in the aqueous solution of metal salts. A schematic diagram illustrating this procedure is shown in Figure 6.5. A dc source is connected between the substrate (cathode) and the metal to be deposited (anode). The deposition rate can be controlled by the current density in the bath. Electroless process is a technique for deposition through the catalytic action of the deposit itself, without the use of a source of external current. Electroless baths for the deposition of gold, copper, platinum, nickel, and palladium are available commercially.

In general terms, the plating process is inferior to those of vacuum deposition. High film purity is hard to achieve, the surface topography is relatively poor, and thickness control is harder to achieve. For these reasons, plating is more generally used in a supporting role to vacuum

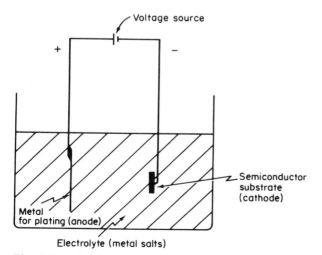

Fig. 6.5 A simple diagram illustrating the plating process.

104   Metallization, Interconnections, and Packaging

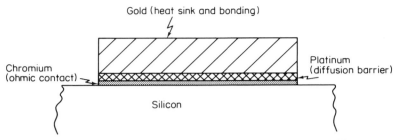

Fig. 6.6 The metabolization system for an ohmic contact to silicon.

film deposition. Hence, the basic metallizations are deposited by vacuum deposition and plating is used to thicken the metallization. (A good example is the production of a thick metal heat sink, a very important feature of high-power devices, as is shown in the Impatt diode of Figure 6.6.)

6.4 Metallization System

For convenience, the metallization systems used in microtechnology fabrication may be divided into functions:

(i) Metals that perform specific electrical interaction with the semiconductor [i.e., such as ohmic or rectifying (Schottky) contacts].
(ii) Those used to interconnect the various circuit elements in an integrated circuit.

In group (i), the main constraint is to ensure that the metal semiconductor contact has the correct electrical properties: ohmic or rectifying. In principle, this can be achieved by matching the metal work function to that of the semiconductor.[1] In practice, however, the existence of a relatively high surface state density in both silicon and gallium arsenide precludes this possibility. Most of the noble and refractory metals result in Schottky barriers and rectification. To fabricate ohmic contacts, the semiconductor just beneath the metal must be doped degenerately (i.e., with a doping of 10^{19} to 10^{20} atoms cm^{-3} or more). Metal contacts to highly doped semiconductors always produce nonrectifying contacts because, even though barrier layers are produced, the depletion layers so formed are sufficiently thin to enable electrons to tunnel through.[2]

---

[1]For full discussion of the physics of barriers, see Appendix 6, S. M. Sze, reference 6, p. 248 or Rhoderick, reference 7.
[2]See reference 6, p. 248.

6.4 Metallization System

One way of achieving these localized highly doped layers is by alloying certain metal contacts. This involves heating the metal/semiconductor system up to the 300° to 600° C range in an inert atmosphere to allow the metal and semiconductor to intermix at the surface. In *n*-type GaAs, for example, ohmic contacts may be fabricated by depositing a mixture of 88 percent gold and 12 percent germanium as the contact metal and when this is heated to 450° C for about 1 min, a good ohmic contact results (Figure 6.7*a*). For silicon, the near contact degeneracy is achieved by using ion implantation doping at low energies (Figure 6.7*b*) or by predeposition from a diffusion source. Such high surface doping levels cannot be attained in gallium arsenide, and the thermal alloying process described earlier is necessary. For metallizations in general, there are other reasons for the choice. These include:

a. A good adherence to the semiconductor.
b. The requirement that the metal layer and substrate be capable of being selectively etched.
c. The ease with which wires can be bonded.

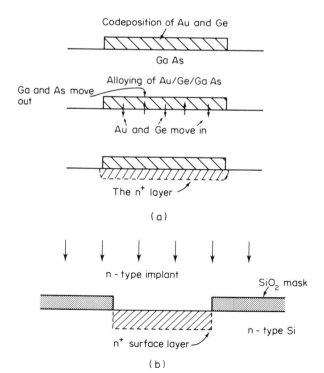

Fig. 6.7 The production of metal/n^+ ohmic contacts by (*a*) alloying Au/Ge to GaAs and (*b*) n^+ implantation into silicon.

d. A resistance to degrading during operation at high current levels and elevated temperatures.

In this context, problems can occur from thermally induced chemical reactions or interdiffusion and electromigration. In some devices, no single metal can satisfy all of these requirements and multilayer metallizations have been developed. As an example, let us consider the high-powered silicon Impatt diode shown in Figure 6.6. Here, the thin chromium layer is used as the ohmic contact with good adherence. The platinum is used as a stable barrier metal to stop the outer gold contact diffusing through the chromium into the silicon device, where it is known to cause rapid degradation at operating temperatures in excess of 300° C. Gold is used for the upper metal partly because its ductile, oxide-free, nature allows easy bonding, and because it is highly resistant to chemical attack and from oxidation during operation in oxygen. It must, however, be emphasized that it is always desirable to use the simplest, least expensive metallization. A compromise must be made between performance and reliability on the one hand and system cost on the other.

6.5 Surface Protection and Wafer Thinning

At this stage in the processing, the microcircuit is a complete entity, ready to be interfaced with the external circuit. There are three small but important steps to be carried out before proceeding. First, the entire active surface area of the wafer needs some form of protection from physical damage (i.e., scratching) or from chemical contamination from its working environment. A good way of achieving these objectives is to cover the entire wafer surface with a strong, impervious layer like silicon dioxide. Silicon dioxide will act as a hermetic seal for all of the devices and interconnect metallizations. Such layers can readily be formed by the sputtering process described earlier in this chapter. But silicon dioxide is not the only possibility; there is considerable interest in silicon nitride as a passivation dielectric, particularly when working with gallium arsenide. An important point to note is that holes will have to be cut above the metal contact pads to facilitate making contact with bond wires and the package the device is mounted in. This can be done using photolithography.

The second problem arises from the finite thickness of the semiconductor wafer. When handling and working with the wafers, they are subjected to stresses and have to be reasonably thick ($\geqslant 150$ μm) to avoid breakage. For the final IC (or single device), a thick layer of semicon-

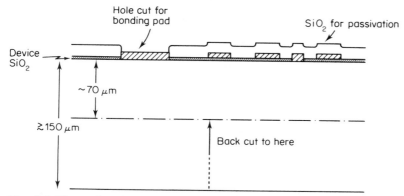

Fig. 6.8 Illustrating the back cut (thinning) and glass passivation of an IC.

ductor lying between the surface where all the action takes place and the metal surface to which it is to be attached is undesirable. The reason for this is that any heat generated during the operation has to be extracted in this direction. A thin substrate enables a more efficient thermal conduction path to be achieved. Other considerations are the following:

a. The dicing process to be discussed in the next section is easier when dealing with thin wafers.
b. If diffusion has been used during processing, the back surface will contain unwanted diffused layers, which could interfere with its operation.

These difficulties are removed by starting with thick layers and later lapping (removal by mechanical polishing) the rear surface, as is shown in Figure 6.8.

In instances where the objective is to mount the chip into a package using an eutectic process, it is desirable to cover the rear surface with a metal layer, such as gold, using a metal deposition technique.

6.6 Dicing, Mounting, and Bonding

The previous sections have been concerned with interconnecting a range of circuit elements situated in a planar configuration over the semiconductor. Metallization processes of this type are very cost effective and can be controlled with excellent precision and, consequently, result in high reliability. A microdevice or microcircuit produced in this way must now be interconnected to much cruder and physically larger electronic components. Consequently, the device or integrated circuit has to be cut from the wafer and mounted in a suitable package; very fine bond wires then must be used to connect the package electrical terminals to those on the chip.

In the first stage, the device or circuit chip has to be separated into its individual components; this may be a single device like an Impatt diode for microwave work as was discussed previously or, alternatively, a self-contained integrated circuit containing many thousands of interconnected components (Figure 6.9). This process, termed "wafer scribing," can be achieved in many ways. Perhaps the most obvious is to use a diamond impregnated circular blade to cut up the wafer (Figure 6.10a). This technique is attractive, since it lends itself to automation at the factory level; large banks of blades are mounted in parallel and are correctly spaced so that they can simultaneously cut along the entire length of the wafer. One problem with sawing is the loss of material resulting from the finite width of the blade. This can be minimized by the use of thin saw blades.

A second method is that of cleaving the crystal along one of the major planes in the structure. When using this process, the devices (or ICs) are laid down in such a way that their periphery squares are lined up with the known cleavage planes. Thus, when the circuit is complete, a diamond tipped scribing tool is drawn precisely across the wafer along the desired breakage line (Figure 6.10a). This is subsequently repeated over the whole wafer. To break the wafer, it is bent on both sides of the scribe line and breakage occurs along the imperfection. As an alternative to scribing, a laser may be used. A laser is pulsed to produce a sequence of small holes along the break line, and this technique works in a manner similar to the scribing technique with the holes functioning as the break defect (Figure 6.10). Because of the condensation of material in a crater around the hole (i.e., called kerf), this technique is best carried out along the rear surface where kerf will not damage the IC components.

Once the individual chips of semiconductor are separated by one of the processes described above, they must be mounted into a convenient package so that the customer can handle and install such a device or IC in his circuit (if left in chip form, it would require expensive equipment and a specialist trained to use such integrated circuits). The chips may be mounted in packages by attaching them to either an epoxy resin, or preform, of eutectic "die-attach." The epoxy is simply a glue that is placed between the rear surface of the chip and the package pedestal.[3]

---

[3]Conductive or nonconductive varieties of glue can be used, depending on whether the back surface is used as an electrical contact.

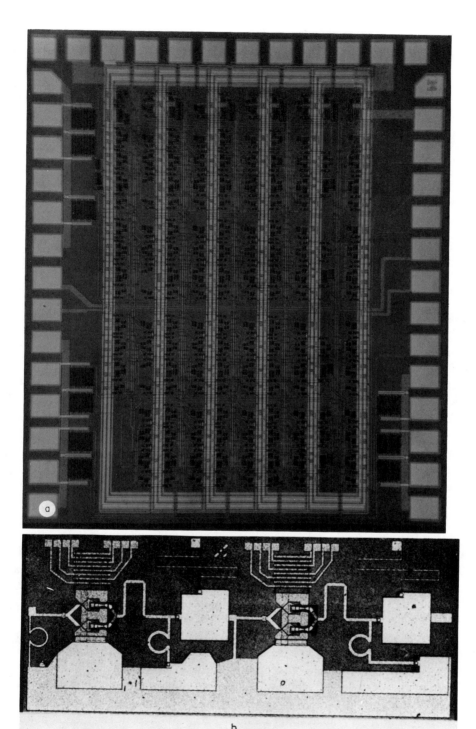

Fig. 6.9 (a) A photograph of an IC, in this case a 5 × 5 multiplier in a GaAs digital IC (permission of Rockwell International); (b) GaAs microwave integrated circuit (permission of Plessey Company).

110 Metallization, Interconnections, and Packaging

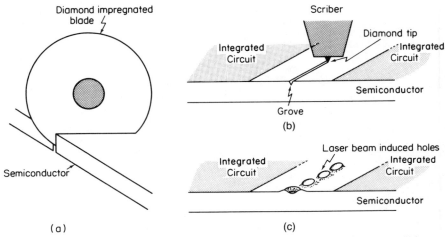

Fig. 6.10 Dicing of ICs: (a) diamond saw; (b) diamond scribe; (c) laser scribing.

Once these epoxy glues are baked to cure them, they provide good adhesion.

In a similar manner, preform bonding is achieved by placing a special piece of metal (called the preform) between the semiconductor and package. This metal, when heated, melts at a low temperature and subsequently, when cooled, will stick to both the semiconductor and package. The third possibility involves an initial deposition of a thin layer of metal (such as gold on silicon) on the rear side of the chip. When heated above the eutectic temperature ($\cong 370°$ C for gold-silicon), the eutectic forms the required bond.

The final step in completing the device is to provide the fine wire bonds that go from the final package terminals to the microscopic metal contact pads on the device (or IC) chip. This process is very labor intensive and demands skill, since it must be carried out under a microscope. The concept is simple, a fine (25 μm diameter) wire has to be attached to the package and to the device, as is illustrated in the example in Figure 6.11. This can be done either by thermocompression or ultrasonic bonding.

The first of these techniques, as its name implies, uses a combination of heat and pressure to secure a firm mechanical and electrical contact between the two metal surfaces (wire and pad on the chip and wire and post on the package). A schematic diagram showing the basic steps in the bonding sequence is shown in Figure 6.12. The end of the wire is melted so as to form a sphere under the influence of surface tension forces. The integrated circuit is heated to around 220° to 260° C and the ball is forced down onto the desired pad. Under the combined effect of

6.6 Dicing, Mounting, and Bonding   111

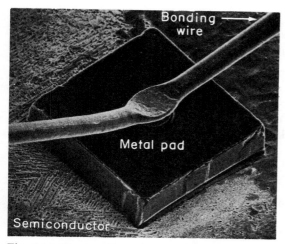

Fig. 6.11 The bonding wire attachment to a low-power microwave-transferred electron device (permission of Plessey Company).

the heat and the pressure, the flattened metallic ball is forced into intimate contact with the pad and adheres to it as the tool is retracted, as is shown in Figure 6.12d. The tool is then taken over to the package contact and the same procedure is repeated; here, however, there is no ball to compress, and the pressure is exerted out to a section of the wire, as is illustrated in Figure 6.12e. The heating is removed and the wire is cut. An example of a device bond is shown in Figure 6.11, which corresponds to a microwave transferred electron device. The whole of the wire bonding procedure is carried out in special equipment with all of the operations performed by observing through a microscope.

The second bonding technique, ultrasonic bonding, is in many ways similar to thermocompression bonding in the sense that the two surfaces

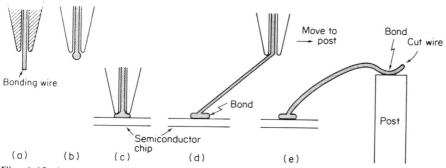

Fig. 6.12 A sequence of curves showing the thermocompression bonding process.

are again forced together by the use of a special bonding tool (Figure 6.12). In this technique, the package is not heated, but the bonding is subjected to ultrasonic vibration to compress the surface together to achieve bonding. An important use of this technique is in situations where it is undesirable to heat the semiconductor chip. Figure 6.13 shows an example of a special bonding tool used for this work. Note how the bond wire is fed through a side hole to the base of the tool where the bond occurs. The usual wire used for ultrasonic bonding on aluminum pads is 99 percent aluminum and 1 percent silicon, but gold wire is reasonably successful on aluminum pads and very good when bonding to gold pads. These two bonding techniques can be combined in the so-called thermasonic bonding. In essence, it is an ultrasonic bonder with limited heating of the package (i.e., $\leq 150°$ C). This is particularly valuable in situations where the chip cannot take the higher temperatures required for thermocompression bonding.

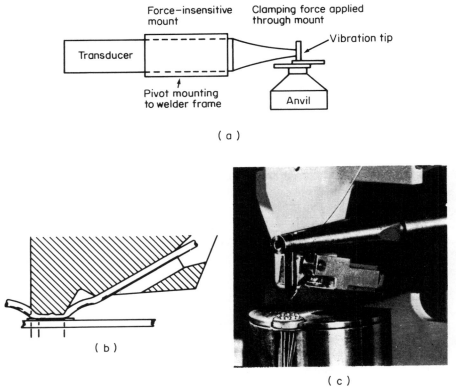

Fig. 6.13 The ultrasonic bonding process: (*a*) schematic diagram; (*b*) the bonding tip; (*c*) photograph of the tip.

6.6 Dicing, Mounting, and Bonding   113

One very important point that emerges from the preceding discussion and relates to production costs is that the microtechnology fabrication processes are divided into two kinds. On the one hand, process steps allow large numbers of identical ICs (or discrete devices) to be fabricated in the same sequence of process steps; a single wafer may contain many hundred circuits at one time. Furthermore, they rely on proven technologies that are readily automated. This is true of the metallization, surface passivation, ion implantation, and diffusion processes—indeed, all of the component technologies for the integrated circuit proper. From an industrial point of view, the natural consequence of this is to work with larger diameter wafers, thus increasing the number of given IC units per fabrication cycle (and hence, per dollar). With silicon processing, for example, the standard wafer diameter has increased from the 2 in. diameter of a few years ago to the present 6-in. diameter. The second class of processing involves items such as dicing the wafers, mounting and packaging, and bonding and testing. These processes are labor intensive because they are not amenable to batch processing. They are, consequently, relatively costly stages. This does, however, illustrate why an IC containing 1000 or more transistors plus a similar number of passive components is so much cheaper to produce than the equivalent discrete circuit.

6.6.1 FLIP-CHIP AND BEAM-LEAD BONDING

The wire bonding process discussed in the preceding section is not always the best engineering solution to the problem and, consequently, special alternatives have been developed. The "flip-chip" technique, as its name suggests, is the procedure where a diode is mounted in a package in an upside down fashion (i.e., with the device or IV ship/substrate facing upwards). There are two reasons why this may be desirable. In some device structures, such as the microwave transferred electron devices (TEDs) or impact avalanche transit time (IMPATTs), the power density in the active region of the device can be as high as 10^{14} W m^{-3}. To remove this heat, the diode must be placed on an efficient heat sink. If an IMPATT diode is flip-chip mounted, as is shown in Figure 6.14a, the active region of the device is right next to the metal heat sink, thus removing the need for the heat to be conducted away through a relatively thick substrate.

In the case of an integrated circuit, the bond wires described earlier result in wasted space and parasitic stray inductances. Flip-chip mounting of such chips can partly overcome these problems. Figure 6.14b shows how an IC chip is bonded to a substrate containing a predetermined

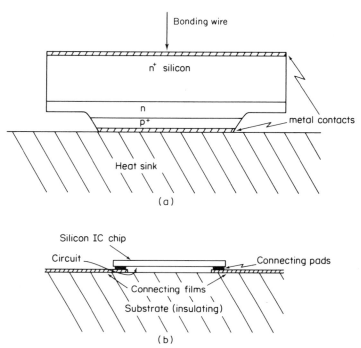

Fig. 6.14 Illustrating the flip-chip process for (*a*) a single IMPATT device and (*b*) an IC chip.

pattern of connecting wires on its surface. Note that raised bumps are attached to the chip, enabling it to make contacts only at specified points.

The beam-lead process is an alternative to the "flip-chip" process and is very valuable in circumstances where stray parasitic elements must be minimized. This is very important in high frequency operations, for example, in microwave devices. The process is to form gold beam leads

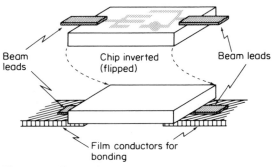

Fig. 6.15 Illustrating the beam-lead process.

on to the chip while it is part of a larger silicon slice. Later, the individual chips are etched out (i.e., not diced or sawn), leaving the "beam leads" overhanging the edge of the chip, as is shown in Figure 6.15. The chip is then inverted (flipped) and mounted onto a substrate in much the same way as the flip-chip example described above.

Problems

6.1 Explain the differences between vacuum deposition and electrolytic metallization techniques.
6.2 What is meant by the term "sputtering"?
6.3 Explain the gold-platinum-chromium ohmic contact system for silicon.
6.4 What happens if refractory metals are used to make contacts on silicon?
6.5 What type of metals are used to make ohmic contacts on GaAs?
6.6 What is meant by the term "passivation"?
6.7 Why is wafer thinning desirable for ICs?
6.8 Explain the advantages of the flip-chip bonding system.
6.9 Explain the disadvantages of the flip-chip bonding system.
6.10 Explain the advantages of the beam-lead bonding system.
6.11 Explain the disadvantages of the beam-lead bonding system.

CHAPTER 7

Fabrication of Devices and Circuit Components

Instructional Objectives

This chapter presents how basic circuit components may be fabricated. After reading this chapter you will be able to:

a. Describe how Mesa diodes are made.
b. Describe a planar diode.
c. Describe planar transistor structures.
d. Describe a junction field effect transistor structure.
e. Describe a metal semiconductor field effect transistor structure.
f. Describe metal insulator semiconductor structure.
g. Describe integrated resistors and capacitors.

Self-Evaluation Questions

Watch for the answers to these questions as you read the chapter. They will help point out the important ideas presented.

a. Explain the difference between a Mesa and a planar structure.
b. What is considered the optimum technology?
c. Why is a buried layer sometimes used?
d. Explain what "field effect" means.

e. Explain the difference between enhancement mode and depletion mode.
f. What is meant by the term "transferred electron device"?

7.1 Introduction

This chapter links together the component technologies described in the five preceding chapters to illustrate how the different basic circuit components may be fabricated. To highlight the basic principles involved, our discussion here is confined to simple components.[1] Extensions of these building blocks to complete integrated circuits (ICs) are examined in Chapter 8.

7.2 Simple *PN* Junctions

By way of a simple starting example, consider the fabrication of a simple *pn* junction diode. There are two possible ways to proceed, depending on whether one is concerned with discrete components or components in an integrated circuit (nonplanar or planar technology).

7.2.1 MESA-ETCHED DIODES

Let us first look at fabricating a large number of discrete *pn* junction devices. A reasonable starting point would be a slice of *n*-type silicon grown on a heavily doped n^+ substrate (Figure 7.1). A heavily doped substrate enables good quality ohmic contacts to be made to the lower surface. The first step in the process is to produce a thin p^+ surface all over the sample. This may be achieved by a shallow boron diffusion or, as is shown in Figure 7.1, by a uniform sheet implantation of boron. At this point, photolithography must be used to define the areas for the uppermost metal contacts. A uniform layer of photoresist is spun on the sample and is baked in the manner discussed earlier. A photographic mask comprised of circular dots is used to expose areas of photoresist to the ultraviolet light which, after developing, produces the circular holes in the photoresist. The ohmic contact is then made to the p^+ region by vacuum deposition of a suitable metal (or alloy) over the entire surface of the slice. When the sample is immersed in an acetone solution, the

---

[1] To clarify some of the features of device structures, the vertical scale is enlarged with respect to the horizontal.

7.2 Simple PN Junctions

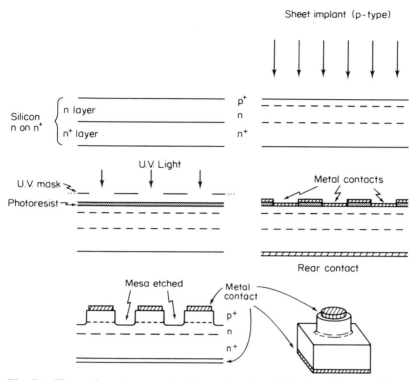

Fig. 7.1 Illustrating the stages in the production of discrete Mesa *pn* diodes.

photoresist covering will be attacked and the metal overlying the resist will *lift off*, leaving intact only the metal ohmic contacts made directly onto the silicon. Success with "lift off" depends critically on the existence of the weak region around the periphery of the hole in the photoresist.

To improve the diode quality, it is sometimes desirable to Mesa-etch, to define and remove any damage (to be produced later by dicing) from the junction region. To do this, photoresist is again spun on and, again using a photographic mask with circular dots slightly larger than the first set but with the centers concentric with the original circular contacts, the regions outside these dots are exposed and removed by developing.[2] With the metal dots thus protected, the semiconductor may be etched away down through the *pn* junction to produce a Mesa, as is shown in Figure 7.1. All that remains now is to coat the reverse n^+ layer with an ohmic contact, to scribe the slice and break it into its single (separate)

---

[2]Alternatively, negative photoresist with the reversed photographic images could have been used.

components, and to mount and bond these components in a suitable package. The two ohmic contacts may require an alloying process or a contact implantation to improve their quality. This depends very much on how heavily doped the initial n^+ and p^+ regions are.

7.2.2 PLANAR DIODES FOR MONOLITHIC CIRCUITS

As a second example, consider a truly planar pn junction. The final device structure is shown in Figure 7.2a, which shows a pn junction formed by successive n- and p-type diffusions (or ion implantations) into a p-type substrate. The natural junction between the n-type diffused region and the substrate provides a good means of electrical isolation between the diode and any other device fabricated on the same substrate, since it must be remembered that in planar technology the diode might be just one component of an IC; Figure 7.2b shows a sequence of events needed to produce any one of these diffused regions. (The photographic mask and expose and develop steps are omitted for clarity.) This sequence of events is carried out first for the deeper n-type diffusion. Then it is repeated with a smaller hole (aligned correctly with respect to the first) for the second p diffusion. In each instance, a new SiO_2 layer must be formed in the manner described in Chapter 4. A third sequence of oxidation/photolithography/etching must be carried out to cut holes in the oxide at the appropriate places to deposit the ohmic metal contacts. Note in Figure 7.2b the steplike nature of the oxide film resulting from the superposition of successive oxide layers in the vicinity of the diode area (omitted in Figure 7.1a for simplicity).

The basic processing steps and device geometry suggested above are not unique, many alternatives are possible. In practice, the optimum technology is the one that minimizes the number of processing steps while guaranteeing a specified device performance. In the case of a complex IC, a trade-off is often required between these conflicting demands. For example, in the diode described above, the device quality could be improved considerably if shallow p^+ and n^+ regions were formed just below the metal contacts to ensure good ohmic behavior. In silicon, this could readily be achieved by ion implantation. However, this would increase the number of processing steps required, which, of course, would be reflected in both processing time (i.e., cost) and in device yield. This latter point should not be ignored since the greater the number of processing steps, the greater the chance of a defect arising in processing and, hence, the lower the *yield* of good components. With regard to diode geometry, the situation is very dependent on the application envisioned

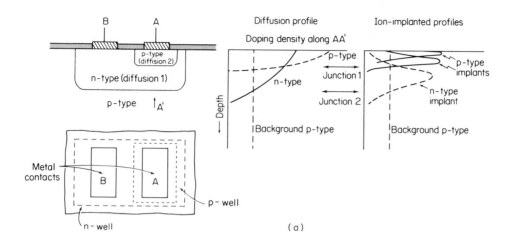

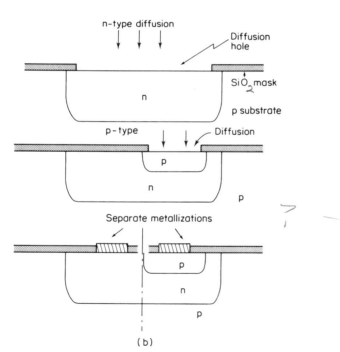

Fig. 7.2 (a) The structure of a planar pn junction diode together with the corresponding diffusion (or ion implantation) profiles; (b) the process steps for the planar diode.

for the diode. Figure 7.3a shows a linear device, but concentric circles would be just as easy as is shown in Figure 7.3b.

As we mentioned previously, a very important point relating to technology is that in any device/IC structure, there are inevitably different ways of arriving at the same end product. The choice of which process to pursue is an engineering design exercise, taking into account parameters such as cost, yield, reliability of the end product, and ease of automation as well as technical merit. An illustration of this involves the device surface metalization with aluminum. This could be used as device interconnections, as in silicon ICs or, alternatively, as a Schottky barrier diode on a GaAs substrate for use as a rectifier or varactor diode. The principles are the same and, to keep the argument simple, they will be illustrated with the latter example. The problem is very basic: a circular aluminum pad is to be deposited onto the surface of an n-type GaAs chip. Figure 7.4 illustrates two different approaches to the problem. Consider first the left-hand branch. Here, the GaAs slice is covered completely with an aluminum film of the required thickness by vacuum deposition. Photoresist is then spun onto the aluminum surface and is baked in the usual manner. At this point, the process branches into two more alternatives, depending on whether one wishes to use positive or negative photoresist. For negative photoresist, the photographic mask must be clear in the area of the circular dots and opaque elsewhere. The developing process would then remove all of the regions unexposed to ultraviolet light. With positive photoresist, the reverse is true and a mask with circular black dots on a clear background would be required. After development, the circular areas of aluminum will be protected by pho-

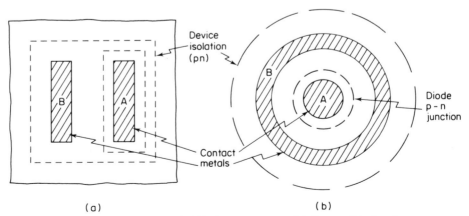

Fig. 7.3 Plan views of diodes of differing geometry: (a) linear; (b) circular.

7.2 Simple PN Junctions

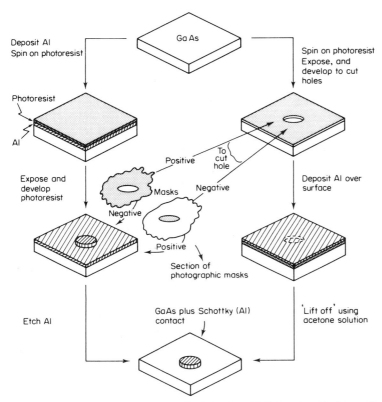

Fig. 7.4 Illustrating two ways of producing Al Schottky diode on GaAs.

toresist. Consequently, when the slice is dipped in an aluminum etch, all of the aluminum not protected by the photoresist will be removed, leaving the desired circular geometry metallization.

The alternative, shown in Figure 7.4, is the lift-off process described earlier. In this process, the semiconductor slice is coated with photoresist and is developed to produce a circular hole in the photoresist film. Here again, there is a choice of positive or negative photoresist with their corresponding masks. The whole system is then coated with aluminum. Finally the photoresist is dissolved away in acetone solvent. The aluminum covering the photoresist breaks away (i.e., lifts off) leaving the required aluminum pads.

The basic processes described above are readily extended to complete IC metallization systems where the geometry is more complex but the basic process steps are identical.

7.3 Bipolar Transistors

The planar diode process described in the preceding section is readily extended to produce bipolar transistor structures. These can be of two kinds, lateral and vertical. Cutaway diagrams of a vertical *npn* and a *pnp* equivalent transistor are shown in Figure 7.5; in both cases, they are fabricated onto a *p*-type substrate. Consider first the *npn* transistor. It is very similar to the isolated *pn* diode described earlier, but here a second diffusion (or implantation) is carried out through a small hole and forms

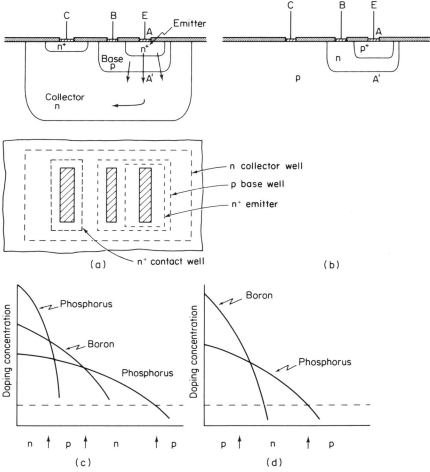

Fig. 7.5 Planar vertical transistor structures: (*a*) n^+pn; (*b*) p^+np; (*c*) and (*d*) the diffusion profiles required, respectively.

the n^+ emitter. A profile along the line AA' would be as shown in Figure 7.5c. To achieve a low resistance contact to the collector, the region under the collector terminal is doped n^+ as shown. The term vertical transistor is appropriate because the major current flow from emitter to base to collector is as shown by the arrows in Figure 7.5a. With a p-type substrate, the pnp transistor is somewhat simpler because the isolation junction is omitted (Figures 7.5b and d). This implies one less diffusion. Note also that, with minor modifications to the diffusion depths and the omission of the collector contact, this device would be just the pn diode described earlier. This raises a very important point. Since every transistor contains two pn junctions, transistors can readily be used as diodes, simply by following the same basic process steps but using different metallization geometries. This is illustrated in Figure 7.6, which shows how the emitter-base junction can be used for rectification. Here, the base and collector are shorted together, thus ensuring that any emitter-base current that flows over to the collector will simply be added to the base current. This allows standard diffusions to be used for both the regular transistors and the diodes. Differences are only required in the plan view geometry and the metallization masks. This drastically reduces the number of process steps needed. There are other possibilities for using the transistor processing to produce diodes. These possibilities are illustrated in Figure 7.6b.

An alternative configuration is the horizontal transistor shown in Figure 7.7. Observe that the base and collector have changed positions and that the current now passes horizontally from emitter to collector. This design appears somewhat simpler in structure but it must be remembered that the base width is now controlled by the photolithographic masks and is, in general, limited to dimensions greater than $\cong 1$ μm.

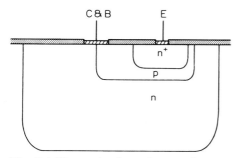

Fig. 7.6 Illustrating how the transistor structure can be used for diodes (in this example the collector-base junction is shorted).

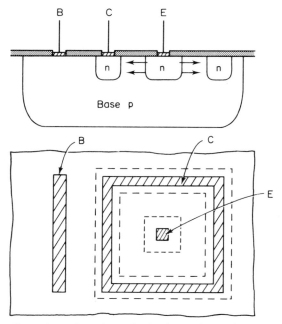

Fig. 7.7 A plan view of a horizontal transistor.

7.3.1 BURIED LAYERS

For simplicity, all the previous transistor diagrams have omitted the buried layer now shown in Figure 7.8. This is in essence an n^+ layer covering almost the entire active region. It has the important effect of reducing the series resistance of the collector while not degrading the collector-emitter breakdown voltage because there is no need to increase the doping in the back of the collector region. In essence, the current takes the shortest path to the high conductivity buried layer, which shorts out the series resistance.

7.4 Junction FET (JFET)

The JFET structure is simply a modified bipolar device as is illustrated in Figure 7.9. The important part of the device is the n-channel formed between the lower isolation junction and the p^+ gate region. The n^+ regions (now shown) are again needed for the ohmic contacts to the source and drain. They also act to ensure that the parasitic resistance in series with the channel are reduced to a minimum.

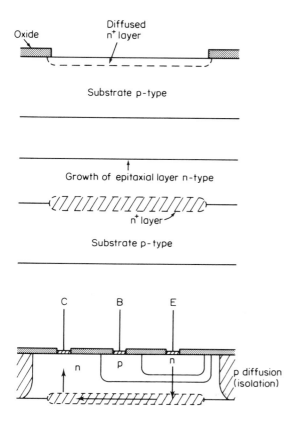

Fig. 7.8 The introduction of the buried n^+ layer into a transistor by epitaxial growth.

7.5 The Metal Semiconductor FET (MESFET)

Another variation on the basic JFET structure is the MESFET shown in Figure 7.10. This is a JFET device in which the junction consists of a metal/semiconductor Schottky contact. It is particularly valuable in situations where *pn* junctions are not easy to form by diffusion (as, for example, in GaAs), and in situations where the junction parasitic capacitance must be kept small (a Schottky barrier exhibits very little diffusion capacitance).

A typical gallium arsenide device structure is shown in Figure 7.10. The active device is constructed on an *n*-type epitaxial layer grown on

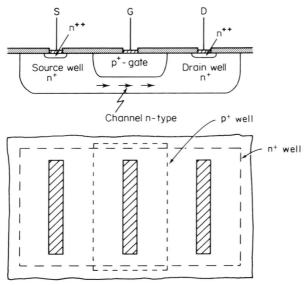

Fig. 7.9 A cross section of a simple JFET structure.

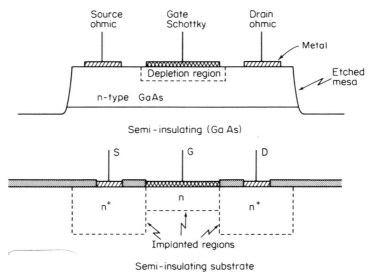

Fig. 7.10 A cross section of a MESFET structure in GaAs.

128

a semi-insulating substrate. This is simply an intrinsic layer which, because of the wide band gap of gallium arsenide, has a low conductivity and, hence, is called semi-insulating. The thickness of the epitaxial layer controls the channel dimensions, since the electrons will not travel into the semi-insulating substrate. The source and drain contacts are ohmic, and the gate is a Schottky diode.

7.6 Metal Oxide Semiconductor Devices (MOS)

MOS transistors are, in essence, field effect transistors that utilize an MOS capacitor structure to control the channel region. Figure 7.11b shows the basic structure of a depletion mode n-channel MOSFET. It is, in fact, very similar to the JFET shown in Figure 7.9 (i.e., the n^+ source and drain regions). Here, however, a thin n-channel extends between the source and drain, at the surface just beneath the silicon oxide layer. A gate metal is situated over the channel region which, with suitable biasing, can modulate the number of electrons in the channel. A positive gate bias will attract more electrons into the channel. Conversely, a negative bias will repel electrons and will pull holes toward the gate. Eventually, this will pinch off in much the same way as the JFET. One of the important advantages of MOSFETs over JFETs is the complete electrical isolation of the gate from the channel. The device shown in Figure 7.11a is an enhancement mode n-MOS transistor (n-MOST). The main difference is that no in-built channel exists and, to turn the device on a threshold, voltage V_T has to be applied to the gate. This is sometimes called the normally off device and is very valuable in logic circuits because it does not consume power when it is not turned on.

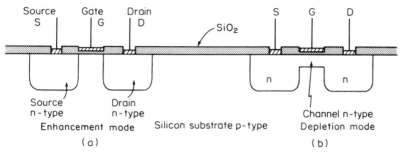

Fig. 7.11 Cross-sectional diagrams of n-channel enhancement and depletion mode MOSFETS: (a) enhancement; (b) depletion mode.

Ion implantation has been described as the new arrival in the field of semiconductor technology. It is fortunate that the technique is ideally suited for planar technology. An excellent example of a situation where implantation provides an appreciable advantage over diffusion is in fabricating high frequency MOS transistors. These devices are normally fabricated by diffusion (Figure 7.12) where the source and the drain are introduced through windows in the oxide mask. Finally, the metal-gate electrode is evaporated so as to span the gap between the source and drain. Because of the limitation to the masking tolerance and the lateral spreading of the diffusing ions, the overlap in the gate and the source or drain can be comparable with the width of the gate. This gives rise to an inter-electrode parasitic capacitance, which limits its high frequency behavior. These capacitances can be particularly troublesome in integrated circuits with their high packing density. Figure 7.12 illustrates how ion implantation can reduce this capacitance effect. In this illustration, the source and drain are produced in the usual way. This gate, however, is considerably smaller than the distance separating the drain from the source.

The complimentary device, the p-channel MOS transistor (p-MOST), is shown in Figure 7.13. It is left as a problem for the student to understand how they may be operated. A combination of the two devices, n-MOST and p-MOST, forms a very valuable unit in logic circuits since, when they are both turned on, the electron current transported by one device can be made to match and neutralize the hole current from the other. Thus, no external current is required and such circuits consume

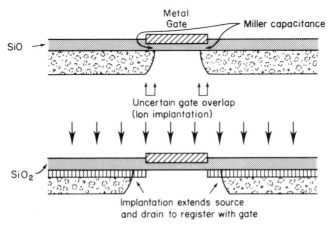

Fig. 7.12 Illustrating how ion implantation can be used to reduce the Miller capacitance of a MOSFET.

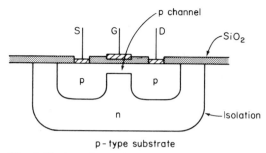
Fig. 7.13 A *p*-channel depletion MOSFET.

low power. The technology is termed complementary MOS or just CMOS for short. The two device structures are compatible and can be run in parallel. CMOS is discussed further in Chapter 8.

7.6.1 VMOS FIELD EFFECT TRANSISTOR

The VMOSFET uses the properties of anisotropic etches to produce this variant with its characteristic V-groove structure, as is shown in Figure 8.5. It is discussed in more detail in Chapter 8.

7.7 Charge Coupled Devices

Charge coupled devices (CCDs) provide a new concept in devices, since they use as their basic parameter electric charge rather than the conventional currents and voltage. A full discussion of their physics and uses is provided in reference 6 of Appendix 6. They are, in essence, delay lines where information in the form of a small charge packet can be injected into an input terminal and transferred in a controlled way along a basic sequence of electrodes. These devices have made dramatic advances since their discovery in the late 1960s, primarily because of the fact that silicon was a suitable material for devices and they could benefit directly from the advanced state of planar silicon technology. A CCD consists of an array of basic cells joined in series. In its simplest form (a three-phase device), each cell comprises three MOS capacitor structures mounted parallel to each other, as is shown in Figure 7.14. By the correct sequence of bias on these three terminals, $\emptyset_1$, $\emptyset_2$, and $\emptyset_3$, a packet of charge in the left-hand-most device can be moved in its entirety to the right into the first MOS device in the next cell. The correct phasing and timing of these bias-pulse by a clock passes the charge packet

132 Fabrication of Devices and Circuit Components

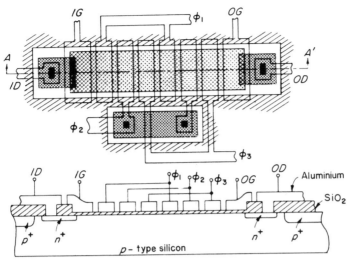

Fig. 7.14 A simple two-cell, three-phase CCD structure.

from one end to the other. The simple structure shown in Figure 7.14 consists of two cells in a buried channel CCD together with the input and output facilities (which will not concern us). The obvious compatibility with the standard silicon system is clear from this figure (i.e., the starting material is a *p*-type epitaxial layer and the active region for the charge transfer is an implanted *n*-type layer). In practical systems, there are many ways of realizing this type of structure, Figure 7.15, for example, shows how the shadow etch technique can be used to etch narrow

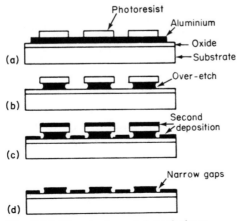

Fig. 7.15 The shadow etch technique.

gaps in a three-phase single-level aluminum gate structure of the type shown in Figure 7.14.

7.8 Passive Circuit Elements: Resistors and Capacitors

Earlier in this chapter, we described how diodes could be obtained by a modification to standard transistor structures; the same is true for some resistive and capacitive elements.

7.8.1 RESISTORS

One of the simplest resistor structures is the diffused resistor shown in Figure 7.16. In this example, an n-type region is diffused or ion implanted into a p-type semiconductor. This layer is electrically isolated by the pn junction and the silicon oxide and carries all of the current between the two terminals. The resistance R can be varied by the geometry (i.e., length L and cross-sectional area A). A simple uniform profile can be obtained by the superposition of implants as is discussed in Chapter 3

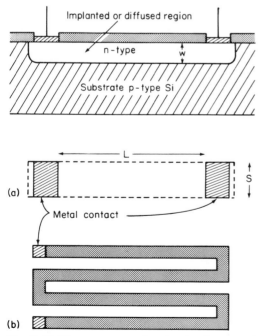

Fig. 7.16 The structure of a simple planar resistor.

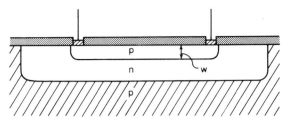

Fig. 7.17 The structure of a simple *p*-resistor.

(Figure 3.21). For large resistor values, the length of the resistor can be economically increased by meandering as is shown in Figure 7.16*b*.

In this simple structure, the control of (*a*) is difficult, and it is fixed by the epitaxial layer thickness ($\cong 8$ μm) rather than by diffusion or ion implantation. This can be overcome by the *p*-resistor shown in Figure 7.17; (*a*) is now controlled by the base diffusion of the vertical BJT and, for such a layer in silicon, a sheet resistance of ~ 150 Ω/square is typical. As a result, resistors of 50 Ω to 10 kΩ can readily be produced in this way.

Another possibility is to use essentially an FET structure (see Figure 7.9). A *p*-type base diffusion limits the cross section of the resistor channel and enables much larger sheet resistances to be obtained and, therefore, higher resistances. The resistor in Figure 7.9 does not have a bias on the gate. If this is metalized so that the gate can be biased, then an electrically variable resistor can be produced. Linear (ohmic) behavior is, however, restricted to the low voltage regions well below saturation, as is shown in Figure 7.18.

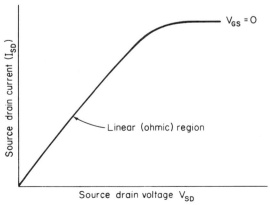

Fig. 7.18

7.8.2 CAPACITORS

The small signal models of transistors (bipolar and MOS) contain many capacitive elements and, in principle, any of these elements can readily be used as a capacitor in an IC. Figure 7.19a shows a typical capacitance voltage curve for a reverse biased *pn* diode. The capacitance falls off with reverse bias. This can be used in devices such as varactors, which are used extensively in tuning electronic oscillators. In a bipolar transistor, two possible junctions may be used. Figure 7.19b, for example, shows an emitter base capacitor (i.e., a n^+p diode), and Figure 7.19c shows a base collector capacitor. One of the problems with this type of capacitor is that the element will draw a finite leakage current in reverse

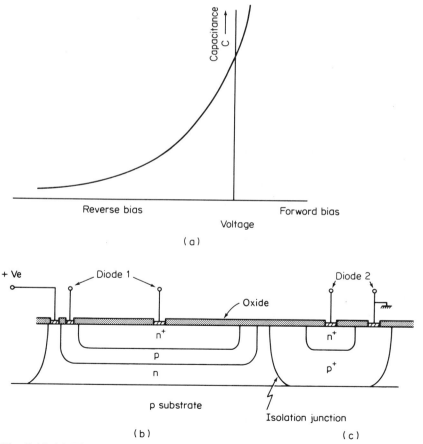

Fig. 7.19 (a) The capacitance-voltage curve for a *pn* junction; (b) and (c) two ways of realizing diodes from planar transistors.

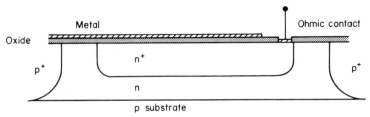
Fig. 7.20 A dielectric MOS capacitor.

bias (i.e., the reverse saturation leakage). In the terminology of the electronic engineer, the quality factor (i.e., Q) of the diode is relatively low. The situation becomes much more difficult if the diode is taken into the forward bias region since, in these circumstances, the series resistance of the junction, which is in parallel with the capacitance, falls to a low value and the diode Q will decrease drastically. For *pn* junctions, the diffusion capacitance will increase rapidly in forward bias, as is shown in Figure 7.19a. A second difficulty is the variations of capacitance with voltage. Ths type of capacitor clearly cannot be used in situations where a constant capacitance is essential.

The difficulties identified above can be removed by using dielectric MOS capacitors. Components of this kind can be produced in silicon by using the surface SiO_2 already on the IC as the capacitor dielectric. It is required to place the dielectric between two highly conductive layers, and this can be achieved in the manner shown in Figure 7.20. An n^+ layer is diffused or implanted into the epitaxial layer and contact is made to it. An aluminum contact is laid down on top of the silicon dioxide to complete the capacitor. The value of the capacitance is varied by a combination of oxide thickness and area.

7.9 Gallium Arsenide ICs

Silicon is the most important semiconductor in use today. Therefore, the IC technology described in the preceding sections is concerned primarily with silicon devices. In principle, all concepts can transfer to gallium arsenide. However, gallium arsenide IC technology is in its infancy, and contemporary research and development indicate that significant modifications are needed before such a transfer can take place. One of the most important differences arises from the lack of a diffusion technology with GaAs, since the surface of GaAs cannot withstand the high temperatures needed for diffusion, as was discussed in Chapter 4. Ion implantation is actively being pursued as a viable alternative and, at this time, can claim considerable success.

A second important difference is that GaAs does not have a native oxide with the dielectric strength or the surface passivating qualities of the SiO_2/Si system. At the present time, no alternative deposited insulating layers, such as SiO_2 or Si_3N_4, have been produced in such a way that the surface state density is sufficiently low to allow surface inversion to be obtained. Metal-insulator semiconductor (MIS) devices, comparable to the silicon MOS devices have not been developed, but active research is being conducted in this direction. The only transistor structure as yet available in GaAs systems is the MESFET device described earlier. This device is equally successful for discrete applications in microwave systems and as an IC element for analog (microwave and multimeter applications) and digital ICs. Future transistor structures using GaAs will be discussed later.

At the present time, two substrate approaches are being used. They are the use of *n*-type epitaxial layers on intrinsic (semi-insulating) substrates and the direct implantation into semi-insulating substrates.

a. *The use of* n-*type epitaxial layers on intrinsic (semi-insulating) substrates (Figure 7.21).* The intrinsic substrates act in the same way as the *p*-substrates in silicon. This only works in GaAs because of its wide band gap. Sometimes a buffer semi-insulating layer is grown to protect the epitaxial layer from the diffusion of impurities from the bulk-grown compensated substrate. Device isolation can be achieved by etching or by proton isolation.

b. *Direct implantation into semi-insulating substrates.* In this process, implantation occurs directly into the semi-insulating substrate with no epitaxial layer. The first implantation is used for the source and drain of the MESFET and then, with a remask cycle, the channel is implanted.

As far as discrete microwave devices are concerned, the devices using epitaxial layers have produced the best performance. They are more perfect than bulk grown material, and consequently have a higher electron mobility and a higher cutoff frequency.

The GaAs IC question is not yet resolved. At this time, the relatively simple microwave and multimeter ICs are indicating clear success. The more complex digital ICs are experiencing some disappointing results partly because of material problems giving low yield. Even with the low yields, GaAs digital ICs are being produced for sale.

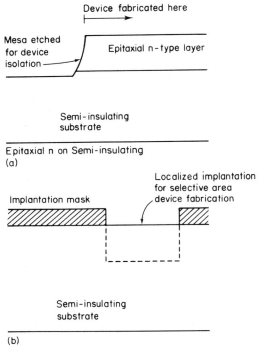

Fig. 7.21 Illustrating two approaches to GaAs IC fabrication: (*a*) Mesa etched epitaxial layers, and (*b*) direct implantation into semi-insulating substrates.

7.10 Special Device Structures

Most of this chapter has dealt with the problems and solutions to IC technology; the simple discrete devices are considered as essentially a very simple IC and follow similar technology steps with occasional special care being needed in bonding and packaging. For example, power devices need special heat sink considerations to prevent the device temperature rising to an unacceptable level. Similarly, microwave devices require very special packages which have low parasitic inductive and capacitive elements associated with them. This is necessary because at the high frequencies of operation, greater than one gigahertz (GHz), the package properties would mask those of the device.

Current semiconductor technology is much more diverse than this simple picture would suggest. There is a bewildering range of device structures that have been developed; some are available commercially but others exist only in research laboratories. We now consider an example that illustrates the direction in which one branch of the current microtechnology is moving.

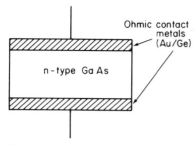

Fig. 7.22 A schematic diagram of a GaAs transferred electron device (TED).

7.10.1 THE TRANSFERRED ELECTRON DEVICE

The basic structure of a transferred electron device is essentially a thin region of n-type material (5 to 10 μm thick for operation in the region 8 to 18 GHz) with doping density $\cong 5 \times 10^{15}$ atoms cm^{-3}, sandwiched between two ohmic contacts (see Figure 7.22 and 7.23). Practical device structures need to be modified in the following two ways:

a. The active n layer is grown epitaxially onto a relatively thick n^+ substrate. Like the MESFET structure discussed previously, it is usual to grow a thin epitaxial buffer layer first to inhibit the outward

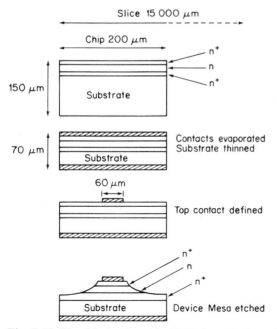

Fig. 7.23 The fabrication of a TED (permission of Plessey Company).

140   Fabrication of Devices and Circuit Components

diffusion of unwanted impurities from the n^+ substrate into the active layer. This substrate achieves two goals:

1. It provides a rigid base on which to grow and process the active epitaxial layer.
2. Good ohmic contacts can be produced on n^+ layers.

b. A thin layer of n^+ is fabricated on top of the epitaxial n layer. This enables good ohmic contacts to be produced at this surface. It is important to note that to reduce the series contact resistances to an acceptable level, both the metal (gold-germanium) contacts must be alloyed at $\cong 450°$ C for about 1 min to increase the near contact (n^+) doping to a degenerate level.

Problems

7.1 Describe a nonplanar integrated circuit technology.
7.2 Describe a planar integrated circuit technology.
7.3 What are the advantages of horizontal transistor structures over vertical structures?
7.4 Draw a cross-sectional view of a vertical *pnp* bipolar transistor.
7.5 Draw a cross-sectional view of a horizontal *pnp* bipolar transistor.
7.6 Draw a top view of the transistor in problem 7.4.
7.7 Draw a top view of the transistor in problem 7.5.
7.8 Draw a cross-sectional view of a *p*-channel JFET.
7.9 Draw a cross-sectional view of a buried *p*-channel depletion-mode MOSFET.
7.10 What is the basic structure of a CCD?
7.11 An integrated circuit resistor is made of pure undoped silicon with a length of 2 mm and a cross-sectional area of 0.01 mm². What is the resistance of this resistor?
7.12 A resistor in an integrated circuit is made using the silicon substrate with a doping concentration N_D of 10^{16} cm^{-3}. If the length is 2 mm and the cross-sectional area is 0.01 mm², what is the resistance of this resistor?
7.13 Describe an MOS capacitor.
7.14 What are the advantages of GaAs over silicon?
7.15 What are the disadvantages of GaAs over silicon?
7.16 What does MESFET stand for?

CHAPTER 8

MOS and Bipolar Technologies and Their Applications

Instructional Objectives

This chapter describes how basic integrated circuits can be fabricated by using both MOS and bipolar techniques. After reading this chapter you will be able to:

a. Explain bipolar technology families.
b. Explain MOS technology families.
c. Describe an integrated circuit design sequence.
d. Describe a bipolar integrated circuit processing sequence.

Self-Evaluation Questions

Watch for the answers to these questions as you read the chapter. They will help point out the important ideas presented.

a. How many diffusions are needed to form a bipolar?
b. How many diffusions are needed to form an MOS transistor?
c. Explain the difference between metal-gate and silicon-gate technology.
d. How is CMOS different from nMOS?
e. How many masks are needed in a typical bipolar fabrication sequence?

8.1 Introduction

The integration of various device structures, such as MOS and bipolar transistors, resistors, and capacitors, are described in Chapter 7. Specific processes that are favored at any given time have emerged, in the fabrication of integrated circuits using these components. Some of them have been superseded as the capabilities of the technology have improved. In this chapter, we first highlight the problem of integrating MOS and bipolar transistors and then present a detailed discussion of MOS and bipolar applications that presently have a dominant role in the semiconductor industry.

8.2 Technology Families

The favored technologies or technology "families" have emerged partly in response to the advances in technology and also in part because of innovative ideas that have arisen. Some have fallen by the wayside because of a particular limitation. A brief list of the more important technologies is shown in Figure 8.1. The major subdivision is based on the type of transistor (MOS or bipolar) used as the active element. The properties of the two types of transistors have an important influence on the relative advantages and disadvantages of the technologies and on how they are implemented. It is instructive, therefore, to review briefly those properties that have a bearing on their fabrication in integrated form. A brief summary of technology families is presented in Table 8.1.

8.2.1 THE BIPOLAR TECHNOLOGY

Figure 8.2a shows the simplest form of bipolar transistor. In this kind of transistor the necessary *npn* structure is usually formed by two diffusions—one *p*-type, the other *n*-type—into an *n*-type substrate. The major current flow is between emitter and collector and is in a direction normal to the surface (usually termed "vertical"). One problem is immediately highlighted by this structure. To integrate many such devices on the wafer surface, all three transistor connections need to be available on that surface. The collector region in Figure 8.2a is not immediately available. This may be facilitated by placing the collector at C' (Figure 8.2a), in which case, the current takes the path shown by the dotted arrow. Additional resistance in the collector circuit is incurred as a pen-

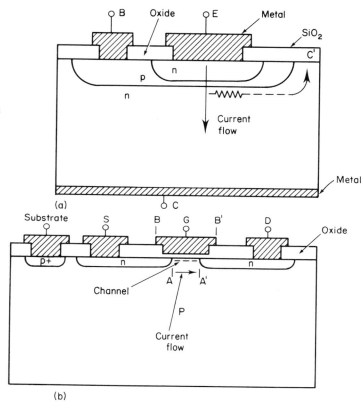

Fig. 8.1 Basic forms of the two transistors used in integrated circuits: (*a*) bipolar; (*b*) MOS.

alty. As we discuss in Section 8.4, the solution to this problem leads to greater process complexity.

A second difficulty with the structure of Figure 8.2a is that if more than one transistor were formed on the surface, they would not be electrically separate. It is necessary, therefore, in bipolar technologies to provide separate isolation for each transistor. This also means greater process complexity and the use of more silicon surface area. A further limitation to the component packing density in bipolar circuits occurs because of the relatively large areas that resistors take up on the surface. The end result is that the circuit packing density in bipolar circuits is not generally as high as for MOS.

However, the bipolar circuit, and in particular ECL, is capable of very

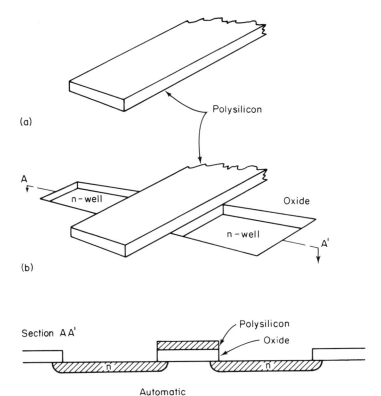

Fig. 8.2 The silicon-gate process: (*a*) polysilicon gate stripe on oxidized silicon; (*b*) source and drain window opened; (*c*) source and drain diffusion.

high-speed operation and is superior in this respect to MOS circuits. Thus, the general conclusion is that the bipolar circuit is used where high-speed performance is important and where the number of components per single-chip circuit is not required to be too high. MOS circuits are capable of very high packing densities but do not have the high-speed capability of the bipolar approach.

An exception to this general conclusion is the I²L bipolar technology. Here, very high component packing densities can be achieved by:

a. Eliminating the need for large-area resistors.
b. "Merging" transistors so that many occupy the area previously consumed by only one.

8.2.2 METAL-GATE nMOS TECHNOLOGY

A simple form of the MOS transistor is shown in Figure 8.2b. Depending on the potential applied to the gate (G), an *n*-type layer is induced in the channel region (hence, the name *n*MOS). This connects together the two *n*-regions forming the source and drain. The resistance of the channel may be controlled by the magnitude of the gate voltage making the device a voltage-controlled resistor. Alternatively, the channel region may exist in either an OFF state, when the induced *n*-region is not present, or an ON state, where a strong *n*-region is present. The device is then a switch and may be used as a binary logic element. The current flow, as indicated by the arrow in Figure 8.2b, is lateral so that the three transistor terminals lie naturally on the same surface in contrast with the bipolar transistor. The whole of the conducting *n*-region (source-channel-drain) is surrounded by a *p*-region. This *pn* junction forms a barrier to the flow of the electrons into the *p*-region. Thus, the device is electrically isolated from other components, and there is no need for separate isolation techniques as in the bipolar device.

8.2.3 SILICON-GATE nMOS TECHNOLOGY

It is essential for the correct functioning of the MOS transistor in Figure 8.2b that the channel region extends the entire way between source and drain, otherwise the current path would be incomplete. Because of the alignment tolerance between masking levels, it is possible that one of the edges of the gate region (B or B') may not overlap the source or drain region (A or A') unless a sufficient margin of error is allowed. Thus, the overlap of the gate over the source and drain regions is greater than is ideally necessary. This leads to larger gate-source and gate-drain parasitic capacitance. The device will, therefore, operate more slowly because those capacitances need to be charged or discharged to change the states, and will occupy more silicon area. This problem led to the development of silicon-gate *n*MOS. This is a variant of the metal-gate *n*MOS device just described in which the gate electrode is formed from polycrystalline silicon. The latter can withstand high temperatures and is present during the source and drain diffusion. The sideways movement of the impurities at the gate edges ensures that the gate region always overlaps the source and the drain. The process and the automatic self-aligning property of the Si-gate process is illustrated in Figure 8.3.

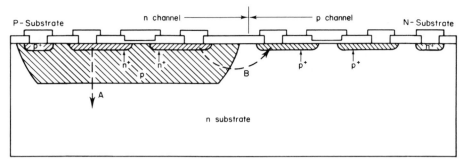

Fig. 8.3 Complementary MOS (CMOS) cell.

Consider a region on a p-type substrate where a gate oxide has been grown. Polycrystalline silicon is deposited and a gate stripe is defined photolithographically (Figure 8.3a). A window stripe is opened in the gate oxide, crossing it at right angles. When the gate oxide is etched, the gate stripe acts as a mask to the etch and two windows are thus opened (Figure 8.3b), one for the source and one for the drain.

When an n-type diffusion is then carried out, the source and drain regions are automatically registered with the gate region, and because of the slight lateral diffusion at the gate oxide edges, the overlap of the gate is ensured. Furthermore, the n-type impurities enter the polycrystalline silicon gate stripe, thereby increasing its conductivity. This, of course, is necessary if the gate stripe is to act as an effective conductor.

The silicon-gate, n-channel process is presently the industry standard and, in the next section, we describe this process in more detail. First, a discussion of the three other MOS technologies that are of some importance in certain special applications. They are CMOS, DMOS, and VMOS.

8.2.3.1 Complementary MOS (CMOS)

In digital circuits, the switching transistors reside in either a "zero" or a "one" state corresponding to it being ON or OFF, respectively (the converse may also be true). In the ON state the transistor conducts current that gives rise to heat dissipation on the chip. This can be significant when the component density is very high. The CMOS technology, which utilizes a p- and n-channel MOS transistor in the switching element, has the advantage that in either the zero or the one state no significant current flows. Therefore, it has very low power dissipation and is used in low

8.2 Technology Families

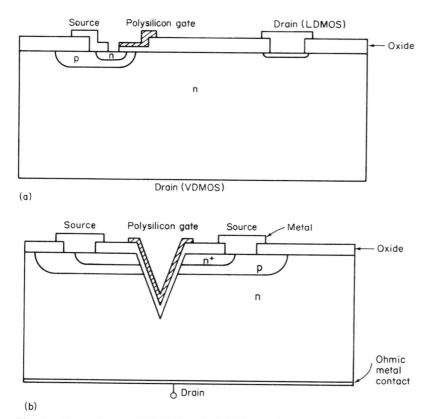

Fig. 8.4 Basic forms of DMOS and VMOS transistors: (*a*) double-diffused MOS transistor (DMOS); (*b*) V-groove MOS transistor (VMOS).

power applications or where heat generation on the chip becomes excessive. The CMOS cell is illustrated in Figure 8.4. The *n*-channel transistor has to have *p*-type semiconductor under the gate and is usually formed by a localized ion implantation or diffusion.

The drawbacks of this technology are that there are more process steps than for *n*MOS and more silicon area is consumed because of the *p*-well. An additional problem is that there are parasitic bipolars (e.g., *npn*; see arrow A in Figure 8.4) and four-layer *npnp* structures (e.g., arrow B), which can give rise to spurious switching or "latching" unless these are designed carefully.

8.2.3.2 DMOS and VMOS Technologies

These two technologies are specialized and have applications in high-voltage MOS devices. Figure 8.5a illustrates a DMOS transistor where

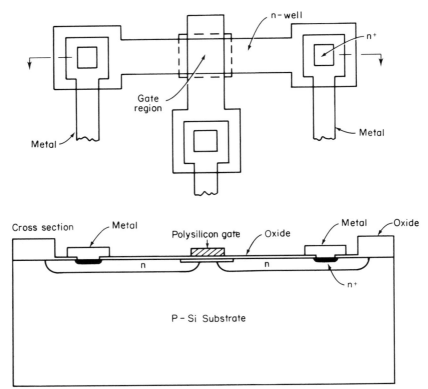

Fig. 8.5 Schematic diagram of Si-gate NMOS transistor.

the channel is defined by a double diffusion. It is possible to achieve very short channel lengths by this technique and to absorb moderate voltages (up to several hundred) across the reverse biased drain region.

In the VMOS structure shown in Figure 8.5b, MOS transistors have been formed on the two faces of a V-groove (see Section 2.4). The channel length is again defined by differential diffusion, and high packing densities are in principle possible. A severe problem with this technology is the precise, uniform etching of the V-grooves over large areas of silicon.

8.3 The Silicon-Gate nMOS Process

In this section, the Si-gate nMOS process is described in more detail, and a discussion of the processing of the transistor with its connections is shown in Figure 8.6. Of course, in a real situation, this kind of a transistor would be only a very small part of the entire circuit and the

8.3 The Silicon-Gate nMOS Process

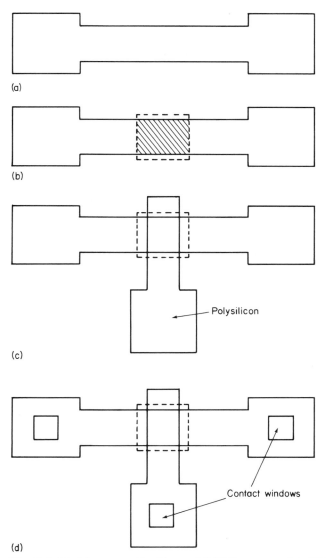

Fig. 8.6 Masking steps for Si-gate NMOS process: (*a*) diffusion opening; (*b*) gate implant; (*c*) polysilicon gate; (*d*) contact windows.

circuit would be one of many on the mask set. Each of the stages that we describe would actually occur simultaneously over the entire circuit and over the array.

a. A thick (typically >1 μm) field oxide is grown over the entire wafer using a wet thermal process. The first mask (Figure 8.6*a*) opens a region where the source and drain will be diffused.

b. It is necessary in some circumstances to adjust the gate voltage at which the MOS transistor turns on. This is done by the gate implant shown in Figure 8.6b. The wafer is coated with photoresist and windows are opened and aligned as shown with the first mask pattern. The photoresist layer itself acts as a mask to the implanted ions. Therefore, the only area that receives the implant is the one that is common to both masks (shown shaded).

c. A thin gate oxide is grown in oxygen over the whole surface, and polycrystalline silicon is deposited. The surface is then covered with photoresist, and the pattern of Figure 8.6c is defined such that the resist is removed from everywhere except within its boundary. The polysilicon layer is now etched away leaving only those regions protected by the resist.

d. The wafer is now diffused with an n-type impurity to form the source and the drain and to dope the polysilicon, hence, increasing its conductivity. The fourth mask is then used to make openings through the oxide for contacts to the source drain and gate regions. These are shown in Figure 8.6d.

e. Finally, metal, usually aluminum or aluminum/silicon, is deposited over the surface, and the fifth mask is used to define the metal pattern to give the final form shown in Figure 8.6a. These electrodes provide interconnections to other parts of the circuit.

Sometimes, before the contact openings are made in step d, the wafer is coated with oxide. This is to give additional protection to the circuit and to minimize the risk of short circuits between the metal and underlying regions.

8.4 The Bipolar Process

The bipolar process described represents some advance from the standard buried collector process invented in the early 1960s. This technology was satisfactory until the circuit complexity exceeded a thousand components per chip. One improvement over the standard process involved the use of oxide as a means of isolating the transistor laterally, leading to a reduction in the area of silicon used for each transistor and, hence, an increase in component density.

The bipolar transistor to be integrated is illustrated in Figure 8.7.

The processing sequence is shown in Figure 8.8, which corresponds to that for an *npn* transistor:

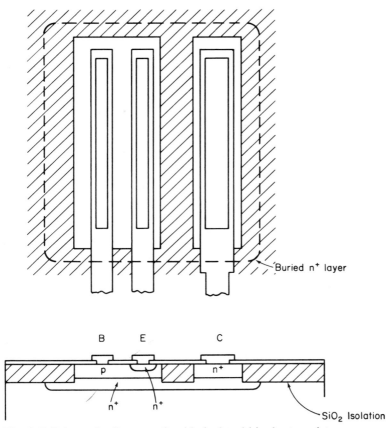

Fig. 8.7 Schematic diagram of oxide-isolated bipolar transistor.

a. A field oxide is grown on a p^- substrate and an opening is made using the first mask for the diffusion of a buried n^+ collector region. The buried layer has two functions: it isolates the transistor from the underlying p^- region by the n^+p^- junction formed, and it gives a highly conducting path for the collector current.

b. A p-type epitaxial layer is grown and a Si_3N_4/SiO_2 composite film is deposited by using CVD. This film is etched away except over those areas that are to form the base and emitter, and the collector regions. The second mask is used for this step.

c. Part of the epitaxial layer in the unprotected areas is etched away and SiO_2 is regrown. It grows very slowly where the nitride film is (see Chapter 4) and much more quickly in the other areas. Thus, it grows downward to join the n^+ buried layer completing the isolation of the

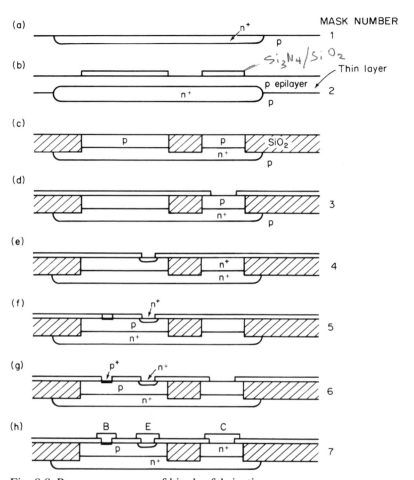

Fig. 8.8 Process sequence of bipolar fabrication.

transistor, and upwards to give an approximately plane surface. A window is now opened over the collector region to diffuse an n^+ contact down to the buried layer (mask 3).

d. An opening is made for the implantation of the n^+ emitter (Figure 8.8e) using the fourth mask, and the emitter region is implanted with phosphorus.

e. Similarly, the base contact mask (number 5) is used to implant p^+ ions in the base contact region. The base itself is the original epitaxial p-layer but, to make a low-resistance contact, it is necessary to increase the p-type density in the vicinity of the contact.

f. Contact windows for base emitter and collector are opened with the sixth mask (see Figure 8.8g).

g. The entire surface is coated with aluminum, and the connection pattern is defined with the seventh and final mask.

The technology just described is not the only one being used in the semiconductor industry. Other variants and improvements are being explored and used by manufacturers in an effort to obtain the best compromise between performance, complexity, and yield.

8.5 Integration of a Commercial Circuit

In this section, we present the application of the technologies described in preceding chapters to the integration of a high performance digital-to-analog converter (DAC) manufactured by the Plessey Company.[1]

We do not give a detailed description of the function of the circuit as this is beyond the scope of this book. Instead, we discuss the sequence of operations the designer followed in the design of this circuit and then highlight some of the design options that the designer had to consider.

Finally, we illustrate where various parts of the circuit actually appear on the chip.

8.5.1 THE DESIGN SEQUENCE

The design sequence is illustrated in Figure 8.9. First, the circuit type must be decided. There are several methods of achieving the D-A conversion, and the choice must be made based on several considerations. First and foremost, will the technique give the performance required by the specification? Second, will the implementation of that technique on the chip give rise to any problems? For example, there may be many more components for one technique as compared with the other, or it may contain components difficult to integrate with the required tolerance.

The technique chosen for the Plessey DAC was the so-called "multiple-current approach." The circuit is illustrated in Figure 8.10. It has the property that if each transistor pair is designed identically and is placed close together on the chip, they will be matched so that if similar care is taken with the resistors R, they will switch equal currents at equal current densities and thus will have closely matched speeds.

---

[1] Plessey Company, Caswell, UK, and Anaheim, Calif.

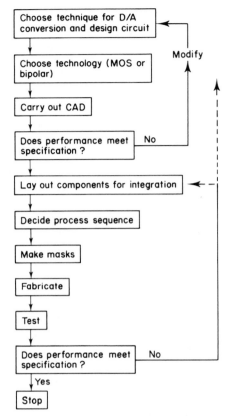

Fig. 8.9 Design sequence for D/A converter.

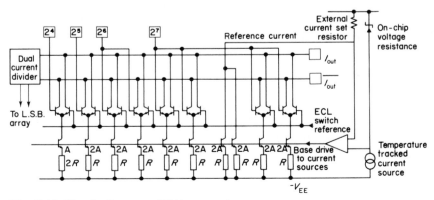

Fig. 8.10 Circuit diagram of D/A converter.

8.5 Integration of a Commercial Circuit

This would not be possible with a more common technique which uses *R-2R* ladder networks, since each stage operates at half the current of the previous stage. This would give rise to different current densities between the different stages so that the temperature differences arising from internal heat generation would change the operating characteristics of the individual stages.

Following the choice of circuit type, a decision has to be made on which technology to use. The first option faced was either to use MOS or bipolar. In the case in question, the component packing density is not very high but the performance is crucial; the primary aim in the design is to produce an eight-bit DAC with a settling time less than 5 ns. This kind of performance would make it attractive to potential users; if the settling time was 100 ns, it would be unlikely to be a worthwhile product. The choice clearly then has to be for bipolar because of its inherently higher speed. A secondary decision to be made is which bipolar technology to use. Here, the demands on speed suggest the use of emitter coupled logic (ECL). In this technology, the transistors do not saturate and, therefore, enable the very low switching times to be achieved. The circuit thus chosen to fulfill the required function is shown in Figure 8.10.

The next question that arises is whether the circuit, as conceived, will function with the required specification. Here, the designer is faced with three options. First, she could simply proceed through the full integration sequence and produce chips with the circuit on which she then could test. This is clearly unrealistic because of the great expense in layout, mask-making, and processing (more so if several iterations are required to meet the performance requirements). Second, she would build the circuit by using discrete components ("breadboarding") and measure its performance. In the present example, the operation of the circuit is very sensitive to the layout on the chip so that the designer would be unlikely to get the required information. Third, computer aided design (CAD) could be used. Over the last few years, software packages have been developed that will calculate the performance of circuits of moderate complexity very swiftly and economically. This approach has the advantage that various parameters can be adjusted to optimize performance. Information may also be obtained on the sensitivity of the performance on small changes in component values that inevitably occur through process variations. In addition, the effects of temperature changes from one device to the next may also be observed. Thus, primary performance may be established, and a great deal of secondary information may be obtained before the circuit is built.

If initial CAD runs show that the circuit has design errors or does not

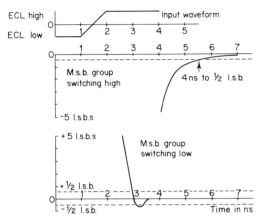

Fig. 8.11 CAD results for current switches.

perform adequately, the designer will have to adjust the circuit design, recalculate its performance, and continue this iteration until satisfactory operation is obtained. In the case of the Plessey design, CAD was used to establish that the circuit of Figure 8.10 operated with sufficient switching speed. The results of the CAD predictions following optimization are shown in Figure 8.11.

The next question facing the designer is how the components of the circuit should be layed out on the chip. Particular care has to be taken so that the transistors and resistors are closely matched and operate at the same temperature. An indication of the layout is given in Figure 8.12. The main resistor array has been placed at the center of the chip and the current switches are placed above and below them.

Following the layout, a detailed process sequence must be decided. A cross section of a transistor used in the circuit is shown in Figure 8.13. A thin epitaxial layer is used with ion implants for isolation, the p^+ regions, base, and emitter. The minimum feature size is 4 μm. The design has similarities with the bipolar technology discussed in Section 8.3, but clearly differs in detail. Resistors were implanted to achieve the close control over their values that is required. We do not describe the detailed processing sequence here because it is similar to that discussed in Section 8.4.

From a knowledge of the layout and the precise geometry of each masking level, pattern generation tapes are produced that contain these data in digital form; they are then used to make the required masks. Finally, the masks are used in the fabrication area, and the circuit is

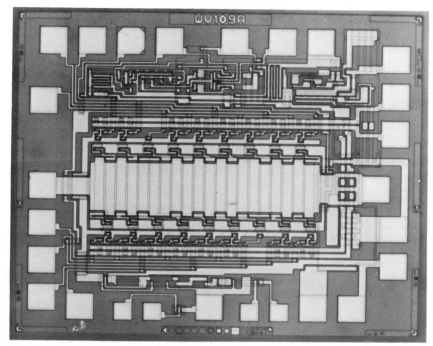

Fig. 8.12 Photograph of DAC chip (courtesy of Plessey Company).

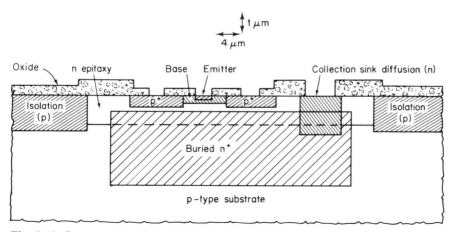

Fig. 8.13 Cross section through bipolar transistor used in the DAC design.

157

TABLE 8.1 **Various Technology Families**

Bipolar	MOS
Transistor-transistor logic (TTL)	Metal-gate p-channel (DMOS)
	Metal-gate n-channel (nMOS)
Emitter-coupled logic (ECL)	Silicon-gate nMOS
Integrated injection logic (I²L)	Silicon-gate CMOS
	Double-diffused MOS (DMOS)

processed in its entirety. A photograph of the complete chip is shown in Figure 8.12.

The measured performance of the Plessey DAC was generally as predicted by the CAD results and fulfilled the performance specification initially aimed for.

Problems

8.1 Describe the operation of a basic MOS n-channel enhancement mode transistor.

8.2 Explain the differences between metal-gate and silicon-gate MOS technology.

8.3 Explain the term "self-aligned gate."

8.4 What are the disadvantages of CMOS over nMOS?

8.5 How many steps are typically involved in designing a simple DAC, as described in Section 8.5?

CHAPTER 9
Future Developments in Semiconductor Technology

Instructional Objectives

This chapter discusses the present state of the art of microtechnology and its future developments. After reading this chapter you will be able to explain what the future developments in semiconductor technology might be.

Self-Evaluation Questions

Watch for the answers to these questions as you read the chapter. They will help point out the import ideas presented.
a. What advantages do integrated circuits achieve over vacuum tubes?
b. What is the projected cost per gate for an integrated circuit in 1990?
c. What are the current operating frequencies for microwave field effect transistors?
d. What semiconductor materials are currently being used at frequencies in excess of 100 GHz?

9.1 Technology: The State of the Art

During the past 20 years there has been an unprecedented development of microelectronics, from the simple diode and transistor structures that displaced the once established thermionic vacuum tubes (or simply vacuum tubes), through simple integrated circuits (ICs), to the staggering complexity and density of current state of the art large-scale integrated circuits (LSI). These circuits currently are reaching levels of 500,000 transistors per silicon chip. The significance of these rather large figures may be hard to fully appreciate for those who have no familiarity with vacuum tubes and their associated power supplies for comparison. It should be stressed that integrated circuits achieve much more than the obvious physical miniaturization of component volume. Two further significant advantages accompany this development: a drastic reduction in power consumption, and an improvement in reliability. These three features cannot be overemphasized. The reduction in power consumption, for example, although not as easy to visualize, is as large as the physical reduction in circuit size. The transistor has no need for the heater power supplies that are central to the operation of vacuum tubes and that represent a 100 percent power loss. The vacuum tubes are, in fact, much worse than this since, in the early vacuum tube operated computers, the power generated for the thermionic process had to be removed to avoid overheating. This, of course, demanded the operation of powerful cooling fans and resulted in further wasted power.

Integrated circuits have two fundamental advantages with respect to reliability. In the first instance, solid state transistors, operating at around room temperature, have a much longer life than that of the heated filament vacuum tubes. This factor alone adds significantly to computer reliability (and would, indeed, to any electronic circuit). However, the current development of the monolithic integrated circuit in preference to solder-interconnected discrete components has resulted in even further and dramatic improvements to reliability. The soldering of component devices is a relatively primitive technology, one very prone to soldering defects (cold-joints), and because it is labor intensive, it is relatively expensive. In contrast, current planar silicon technology is a clean, easily automated, and precise technology. This has led to the large reduction in real cost, an increase in circuit yield, and a trouble-free operation of electronic circuits. This, together with the complexity and versatility of current ICs has resulted in a situation where ICs are now exercising a major influence on our working and leisure life-styles.

One of the most important applications of silicon ICs, the microprocessor, has now influenced most aspects of our leisure and work activities and, indeed, the potential for robotics created by the ICs in industry is, at least for the immediate future, a very worrying prospect for those people who are forced out of employment.

9.1 Technology: The State of the Art   161

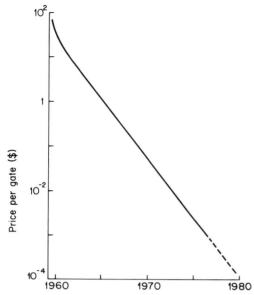

Fig. 9.1 A plot of the cost per gate versus time from 1960 to 1980 (after Roberts, Chapter 1 in reference 5, Appendix 6).

To illustrate the important achievements in cost reduction and reliability, relevant data are presented in Figures 9.1 and 9.2. Figure 9.1 shows the cost per gate versus time from 1960 through to 1980.[1] The logarithmic cost scale shows that the reduction from $90 per gate to ($0.0001) per gate (i.e., a reduction of 1,000,000) in just 20 years is as significant as the reduction in size and power waste.

Looking at the reliability aspects, we become aware of some further information. This is illustrated in Figure 9.2, which shows a schematic curve of the failure rate per function versus the scale of integration. As the scale of integration is increased, there will be an initial fall in the failure rate per function because more of the total circuit is placed on a chip, thus decreasing the number of unreliable interconnections that is associated with an assembly of discrete devices, or small ICs. This improvement does not continue indefinitely, since a point is reached where the complexity is so great that the creation of new reliability problems dominates the circuit performance. One example could be the result of the sheer size of the chip. The larger the chip area, the greater the chance of its containing a destructive structural defect, as was suggested in

---

[1] "Gate" is a term for a logic circuit component consisting of roughly four transistors.

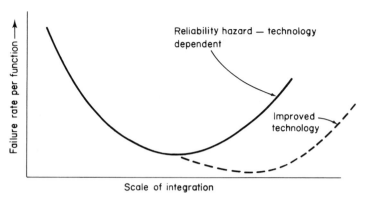

Fig. 9.2 A schematic curve of the failure rate per function versus scale of integration.

Figure 1.1. Consequently, this demands improvements in the growth and preparation of the starting wafers. Another example of this problem is the attempt to increase the scale of integration by reducing the component sizes. At any particular time there will be a limit to technological processes such as the resolution of the lithography. Attempting to push these to their limits will result in a degradation in component quality, and hence, a reliability-yield problem.

These few examples emphasize the fact that the optimum in scale of integration shown in Figure 9.2 is not a permanent limitation but is strongly a function of the technology in question, or at least of its weaker points.

9.2 Future Technology Developments

To the semiconductor industry wishing to ensure a future for itself in the late 1980s and 1990s, the assessment of future developments and trends in semiconductor and IC technology is of prime importance. The rapid and sometimes spectacular developments of the last two decades make this type of prediction very difficult. It must, however, be noted that much of this progress is the direct result of wise investment in research and development, both in terms of manpower and equipment. This trend will continue, but because of the high costs, many of the investment decisions of the future will be outside the domain of the scientist and engineer and in the hands of the government and large industrial corporations.

What then are these technology developments likely to be? Some clearly identified regions are as follows:

a. Higher packing density and greater complexity and VLSI.
b. Higher operating speeds.
c. New semiconductor materials and devices.

The division presented above is somewhat simplified, since many of the objectives overlap. The first of the regions (a) is being studied the most actively today. It is centered on silicon, the "bread and butter" material of the industry. Current development directed at region a will achieve some of the objectives of region b. The technology may translate, with modifications, to any new semiconductors developed in the future (region c).

9.2.1 VERY LARGE SCALE INTEGRATION

An important current development is the drive to very large scale integration (VLSI). Central to this development is the need to decrease component size, with the almost mythical objective of working with device dimensions in the sub-micrometer range. This break point, at around 1 μm, is significant since it represents the approximate limit of conventional optical lithography and, therefore, is the limit of current technology. The two approaches presently being investigated to overcome this limit are electron-beam lithography and X-ray lithography. In both instances, the resolution limit caused by the wavelength of ultraviolet light can be overcome. No industrial semiconductor manufacturer wishing to stay in business can afford to ignore these developments, in spite of the high cost of research.

The sub-micrometer developments will achieve two main objectives: a technology for VLSI, and higher operating speeds for conventional LSI circuits. The higher speeds resulting from the sub-micrometer dimensions are also very important for the development of microwave field effect transistors (i.e., MESFETs or HEMTs). In this case, the small gate width decreases the electron transit time and enables the transistor to operate at a higher frequency. Sub-micrometer dimensions, coupled with the use of the high mobility material gallium arsenide, are currently producing MESFET devices that can operate at frequencies around 40 GHz.

As the sub-micrometer realm is entered, several important difficulties arise. First, there is the problem of scaling. It is not sufficient just to reduce the surface dimension and leave those that are normal to the surface unaltered. Nor can smaller devices be operated at the same voltage. Simple constant field scaling predicts that voltages will drop in proportion with gate-length reduction. However, current densities then increase (as the square of the scaling factor). This leads to a limitation set by electro-migration in the conductors. In addition, if operating voltages are reduced, then noise margins are more difficult to maintain.

The second problem that arises concerns the changes in the physics that occur when devices are made sufficiently small. Impurity atoms are sufficiently closely spaced in present-day devices (e.g., for a doping density of 10^{15} cm^{-3} the mean spacing between dopants is 0.1 μm) so that when ionized, they can be treated as a uniform distribution of charge. However, when device dimensions get to below 1 μm, the granularity of the charges becomes important. Present-day device physics relies heavily on the assumption that electrons and holes are scattered many times during a single pass through a device. Such scattering may occur from other electrons or holes, impurity atoms, crystal imperfections, or photons. Scattering processes are characterized by the average time between individual such events. When device dimensions become sufficiently small, it may be that only a very few collisions will occur during the passage of a carrier through a device. In this circumstance, the basic transport mechanism is no longer scattering dominated but approaches a ballistic-like motion in which carriers accelerate uniformly in the applied field.

Current mathematical descriptions, or models, of semiconductor devices are usually one-dimensional. This is fortunate, since these models describe semiconductor operation accurately, whereas analytic solutions to two- and three-dimensional formulations do not exist, except possibly in a very restricted number of situations where the geometries are artificially simple. However, as device dimensions become sub-micrometer, the simple models are no longer accurate. This means that two-dimensional models must be used and an effort must be made to obtain numerical solutions. The latter usually means very large computation times and a loss of physical insight into the operation of the device.

9.2.2 NEW SEMICONDUCTORS

It is going to be a hard task for any material to displace silicon in terms of cost, availability of base material, and its commanding lead of a high developed planar technology. At the present time, certain III to V semiconductors are being developed for specialized applications where they can improve on silicon. The most important of these is gallium arsenide followed by indium phosphide. Gallium arsenide was initially developed because it exhibits the transferred electron effect and, therefore, is used to fabricate two terminal microwave devices (transferred electron, devices, or just TEDs). GaAs and InP TEDs are still important for providing the basis of microwave power sources up to and in excess of 100 GHz.

Gallium arsenide MESFETs with their high operating speed are now being used as the basis of a new family of integrated circuits. Two specific areas of this work are as follows:

a. Monolithic analog microwave ICs.
b. Digital logic circuits for gigabit operation.

The first of these applications is now well established with a clear role in future microwave systems. The second is still an open question because, although working ICs have been fabricated, they still have not achieved the predicted speed improvement over silicon. At the present time, the major investment in this kind of work is military in origin.

The III to V semiconductors and their ternary and quaternary compounds have a clear role in producing optoelectronic devices, lasers, light emitting diodes (LEDs), photodetectors, and integrated optical systems (the optical sequel to the conventional IC). The rapid development and use of optical communication systems has established a clear role and future for these materials. Long wavelength optical devices (i.e., $\lambda \cong 1.3 \ \mu m$) have enabled current opitical communication systems to operate without repeater stations in links up to 100 km in length, and the move from concept to real systems has taken place in just over 10 years—a remarkable achievement!

9.3 Conclusions

An appraisal of the progress of semiconductor technology thus far indicates that it has been achieved by a combination of continuous technological development, punctuated by discontinuous steps resulting from the discovery of a completely new device. Two very good examples of the latter are the discovery of the transferred electron device mentioned previously and the charge coupled device (CCD) discussed in Chapter 7. The CCDs are very interesting because they link the two types of progress. The concept of using the transfer of a packet of charge rather than current as the basic unit of measure was an original innovation. However, the rapid development and system application of these devices in silicon is a direct result of being able to use the existing planar technology. Herein is a very important lesson: *When we choose substances other than silicon, the development of engineering quality devices is retarded because of the expense and time needed to establish a range of new technologies*. Hence, gallium arsenide and the example of the CCDs made from this material have a long way to progress to match that of present-day silicon CCDs.

Another important problem concerns the design and testing of integrated circuits when the component density is very large. It is estimated that, subject to previous experience and earlier designs, several man-years of effort are required to design circuits that are relatively complex but by no means unachievable by current standards. The projected design costs of circuits that may perhaps be several orders of magnitude more complex become very large and could become prohibitive. Furthermore, there is the problem of testing these circuits once they are made. There is no way that this could be done manually, even at present, and in the future we are likely to see sophisticated automated test equipment evolve that can cope with this differently. Such costs must, of course, be added to the final retail costs of the circuit and may have a profound effect on overall economic decisions.

Finally, there is the problem of need. There must come a point in the advancement in technology where the capability of the circuits match the requirements of the application. A specific example concerns the home computer market. More and more computing power is presently being given to the home computer owner. When systems exist that will fulfill every need, there will be little point in purchasing yet more computing power.

There is no way in which to predict when this will occur nor will it be attempted here. However, an eventual saturation in demand is inevitable in the long term, and this will have perhaps the most profound effect of all on the semiconductor industry, which for the last 20 years (and for an unforeseeable number of years in the future) has been firmly based on a policy of very rapid expansion.

The story of semiconductor technology is not all rosey—there are many blind alleys. In terms of devices, a good example is the tunnel diode. This two-terminal device can operate at microwave frequencies. However, the combination of difficulties in making these devices with reproducible characteristics and the low output power has led to the cessation of their production. It must, however, be remembered that these devices added much to our physical understanding of semiconductors and this can be thought of as an investment in the future rather than wasted effort.

The prediction of future developments in semiconductors is no easier today than it was 15 years ago, and very few people then were able to predict the exciting developments witnessed since that time.

Problems

9.1 What is the cost reduction factor per gate that the industry has seen over the past 20 years?
9.2 What are the technology development areas likely to be in the future?
9.3 What is the device dimension region for VLSI?
9.4 What is the current operating frequency limit for MESFETs?
9.5 What is the current operating frequency limit for TEDs?

APPENDIX **1**

Material and Device Evaluation Techniques

A1.1 Introduction

The semiconductor material prepared for a specific device application requires careful evaluation to see if its parameters lie within the specification for a particular device structure. There are available today a wide range of material and device evaluation techniques with which to characterize both the base material and, in some instances, to evaluate the actual layers within device structures. In this brief summary of the major techniques, our attention will be confined only to the basic principles. We make no attempt to describe the details of practical systems, since some of these are very complex (e.g., SIMS), and most people are usually concerned only with their application.

For device grade semiconductor layers, the following information is usually required:

a. Inspection of layer quality, surface morphology, uniformity, and thickness.
b. Evaluation of layer resistivity (or conductivity) and mobility.
c. Measurement of doping profiles.

Ideally, it is desirable to conduct these measurements on the actual layers being used for the device fabrication. This is not always possible, and sometimes test samples have to be grown alongside the device layer.

A1.1.1 LAYER THICKNESS: ANGLE LAP AND STAIN PROCEDURE

This technique is suitable for estimating the depth of a *pn* junction. The basic idea is illustrated in Figure A1.1 where the junction depth d is small. The wafer is angle lapped by mounting the wafer on a special fixture that allows the edge of the wafer to be wedge polished at a very small angle (1–5°). The small angle taper α enables the distance x to be magnified with respect to d; that is

$$x = d/\sin\alpha >> d$$

knowing x and α, d can be calculated.

To be able to see the *pn* junction, it is delineated by the use of a special etch that will selectively stain either the p or the n region. A useful example for silicon is the chemical etch.

(concentrated hydrofluoric acid and + (0.1 to 0.5)% nitric acid)

which acts to darken (stain) p material.

Note: Hydrofluoric acid is an insidious hazard chemical and care must be taken in its use!

An alternative procedure is to use a scanning electron microscope (SEM) to delineate the junction. This is shown in Figure A1.1 where

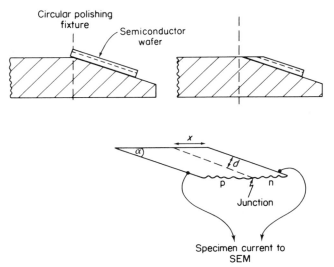

Fig. A1.1 The measurement of junction depth using the angle lap technique. In this example the distance x is measured using an SEM in the specimen current mode.

terminal current across the *pn* junction is only produced when the actual scanned beam is hitting the junction depletion region. In these circumstances the electron-hole pairs generated in the depletion are swept out by the internal field and produce the terminal current. The SEM image is produced from the specimen current (i.e., the so-called specimen current mode of operation).

A1.1.2 MEASUREMENT OF LAYERS SHEET RESISTANCE: FOUR-POINT PROBE

The four-point probe technique can be used to evaluate the sheet resistivity of a layer of semiconductor of known thickness which, in the analysis, is assumed to be homogeneous in its electrical properties. Although it can be used on a homogeneous self-supporting layer, in practice one is usually concerned with a thin near-surface layer that is electrically isolated from the bulk sample. In the case of a thin *p* layer on an *n*-type base (or the alternative configuration, an *n* layer on a *p*-type base), the electrical isolation is provided by the *pn* junction. In the case of gallium arsenide, the *p* or *n* layer is usually grown (or implanted) onto a semi-insulating substrate.

A1.1.3 SPECIFICATION OF SHEET RESISTANCE

Consider the thin sample of material specified in Figure A1.2. If the resistance is measured between surfaces A to B, then resistance can be written as

$$R = \frac{\rho(s)l}{ws} \quad (A1.1)$$

where the specific resistance $\rho(s)$ is assumed to be uniform across the layer and measured in ohms per centimeter.

Equation A1.1 may be written as

$$R = \frac{\rho(s)}{s}\left(\frac{l}{w}\right) = \rho_\square \left(\frac{l}{w}\right) \quad (A1.2)$$

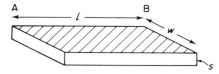

Fig. A1.2

172   Material and Device Evaluation Techniques

where $\rho_\square$ is defined as the sheet resistance. Its units are

$$\frac{\text{ohm-cm}}{\text{cm}} = \text{ohms}$$

Note that for a sample with unit sides, $l = 1$, $w = 1$, and $R = \rho_\square$ regardless of dimensions and, hence, the use of the term ohms per square.

A1.1.4 FOUR-POINT PROBE

The probe is a practical and simple way of determining the sheet resistance. Here, four equally spaced collinear probes are placed onto the layer surface (Figure A1.3a). A current of I (amperes) is passed through the outer probes and a voltage V (volts) is measured by using the two inner probes. Note that these two inner probes do not draw current, and there is no need to worry about any contact resistances. A value of current I is chosen such that V and I are linearly interrelated.

For situations where the layer is p on n or n on p or p (or n) on semi-insulating layer, all current drawn passes in the layer being measured. Now for the outer two probes drawing a current I (Figure A1.3a) the potential at some arbitrary point ρ is:

$$V\rho = \frac{I}{2\pi} \rho_\square \ln\left(\frac{r_2}{r_1}\right) + A \qquad (A1.3)$$

where A is a constant of integration.

Apply this equation to the central electrodes, the four-point probe

$$V_1 = \frac{I \rho_\square}{2\pi} \ln 2 + A$$
$$V_2 = \frac{I \rho_\square}{2\pi} \ln 2 + A \qquad (A1.4)$$

for the potential $V = V_1 - V_2 = I \rho_\square/\pi \ln (2)$.

Rearranging,

$$\rho_\square = \left[\frac{\pi}{\ln (2)}\right] \frac{V}{I} = 4.5324 \ (V/I) \qquad (A1.5)$$

The equation for $\rho_\square$ presented above assumes an infinite wafer. In practice, the use of finite wafers requires a correction factor C_f where C_f is defined by

$$\rho_\square = C_f(V/I) \qquad (A1.6)$$

For the simple case of circular wafers with central probes, the rela-

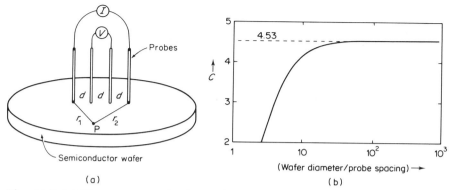

Fig. A1.3 (*a*) The geometry of a four-point probe; (*b*) the correction factor C [equation (A1.6)] plotted as a function of the ratio of wafer diameter to probe spacing.

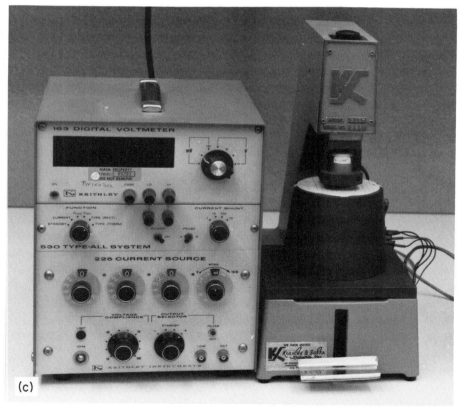

Fig. A1.3 (*c*) Four-point probe (Keithley Type-All System).

tionship between the correction factor C_f and the wafer diameter/probe spacing is shown in Figure A1.3b.

A1.1.5 MOBILITY AND THE HALL EFFECT

The basic principle of the Hall mobility measurement is illustrated in Figure A1.4, which shows the situation for an *n*-type semiconductor. The Lorentz force on the electrons deflects an excess of electrons (positive charge) on the top surface. In equilibrium and internal electric field, F_y is set up to balance the Lorentz force and, hence, the net force in the *y* direction to zero. Generally, $f = -q(\overline{F} + \overline{v} \wedge \overline{B})$, since $V_x B_z$. Furthermore, since $I_x = -nqv_x$,

$$\frac{F_y}{I_x B_z} = \frac{1}{qn} = R_H \quad \text{(Hall coefficient)} \qquad (A1.7)$$

The sign of the Hall voltage determines the nature (electrons or holes) of the charge carriers and their density *n*. Since

$$\frac{I_x}{F_y} = -nq\left(\frac{V_x}{F_y}\right)$$
$$= -nq\mu_H,$$

the mobility μ_H can be determined.

From the point of view of practical measurements, the Van der Pauw clover leaf sample is very convenient to measure the mobility and resistivity of a given layer. The active device layer is cut into the clover leaf shape shown in Figure A1.5. Note that for test samples grown on a semi-insulating (or a blocking *pn*) contact to a substrate, the clover leaf shape

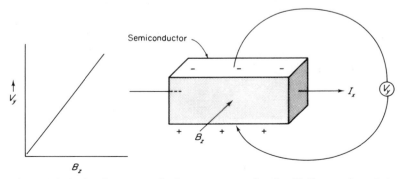

Fig. A1.4 A simple geometrical arrangement for the Hall experiment, together with the Hall voltage V_y plotted as a function of the magnetic field B_z.

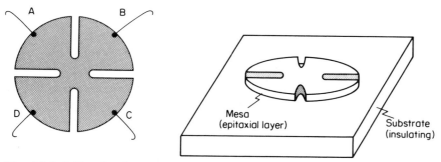

Fig. A1.5 A Van der Pauw sample.

is only cut into the epitaxial layer. With the appropriate masks, this can be done either by Mesa etching or sandblasting. The four contacts at A, B, C, and D are ohmic contacts. If it is assumed that the clover leaf sample is uniform and has only one carrier type, then the Van der Pauw analysis yields

$$\rho = \frac{d\pi}{2 \ln(2)} \left(\frac{V_{DC}}{I_{AB}} + \frac{V_{BC}}{I_{AD}} \right) \qquad (A1.8)$$

$$\mu_H = \frac{d}{B_0 \rho} \left(\frac{V_{CA1}}{I_{BD1}} - \frac{V_{CA0}}{I_{BD0}} \right) \qquad (A1.9)$$

where B_0 is the magnetic field perpendicular to the sample and the extra subscripts 1 and 0 refer to the cases with or without the magnetic field.[1]

A1.1.6 GEOMETRIC MAGNETORESISTANCE

In the Hall effect, the magnetic field was used to set up an electric field that was then used to determine the carrier type, density, and mobility. An alternative procedure, geometric magnetoresistance, can also be used to determine a magnetoresistance mobility μ_m, which is very closely related in value to the Hall mobility μ_H. For this measurement, the active layer geometry must have a large aspect ratio. In its simplest form, a circular sample diameter D and thickness d is placed between two ohmic contacts. Here, the aspect ratio (D/d) must be $\gg 1$. If this layer is now placed inside a magnetic field B, orientated as is shown in Figure A1.6, and the low electric field current density J is determined as a function of B, then the simple relationship[2]

---

[1] L. J. Van der Pauw, Philips Research Reports 13 (1958), pp. 1–9.
[2] For a full discussion of this theory see T. R. Javis and E. F. Johnson, *Solid State Electronics* 13 (1970), p. 181.

176   Material and Device Evaluation Techniques

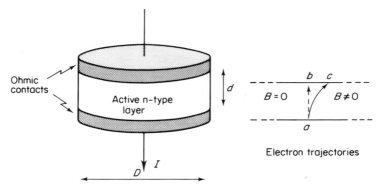

Fig. A1.6 A simple device structure suitable for geometric magnetoresistance (GMR) measurements (i.e. $D/d \gg 1$).

$$\frac{J(B = 0)}{J(B \neq 0)} = 1 + k_s^2 \, \mu_m^2 B^2 \qquad (A1.10)$$

where $k_s^2 \cong 1$. Hence, μ_m can be determined.

In its simplest form, the large aspect ratio inhibits the build up of the Hall field, and the charge carriers are forced to take a longer path (a-c, Figure A1.6) from one electrode to the other; hence, the resistance of the sample increases with increasing B.

One way of using this technique is to grow a special layer along side the layer whose mobility is required and to use a heavily doped n^+ substrate; this helps to ensure a good ohmic contact to the back. Another important feature of this technique is that it can be used to measure the actual mobility of certain device structure. These are transferred electron devices (see Figure 7.23), which have the simple geometry required by

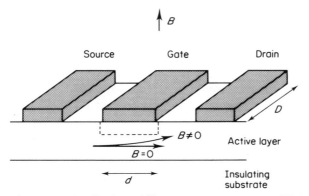

Fig. A1.7 The GMR mobility measurement on an FET device.

A1.6. Typically, $D \cong 200\mu_m$ and $d \cong 10\mu_m$. Similarly, the field effect transistor can also be used to determine μ_m, as is shown for an *n*-type MESFET in Figure A1.7.

A1.1.7 THE MEASUREMENT OF DOPING PROFILES

One of the most important features to emerge from any discussion of semiconductor technology is the importance of the control and the measurement of impurity profiles. They define the device type and control its performance. In very simplistic terms, there are two general types of profiles that may be measured:

a. The electrically active profile.
b. The total impurity profile.

For electrically active profiles, only those impurity atoms that are electrically active contribute (for example, correctly located *p*- or *n*-type dopants and deep-level trapping atoms). In the latter situation, the impurity atoms may be electrically inactive. From the point of view of device performance, the first category is the important one, but information on the second type of profile is frequently needed to determine what percentage of atoms have become electrically active. Some of the most important techniques are summarized in Table A1.1.

In the category (b), the technique provides a measure of the total concentration of dopant (or impurity) in question and does not differentiate between electrically active atoms and others.

A1.1.8 RADIOACTIVE ISOTOPES

This technique is limited to dopants which have a radioactive isotope that has a half-life in a suitable range. The basic principle is as follows: radioactive isotopes are diffused or implanted into the semiconductor. If the total number of atoms is Q, then the use of a suitable nuclear particle detector will enable the degree of radioactivity R to be measured.

TABLE A1.1

Technique	Type of Information
Radioactive isotopes plus stripping	Total (b)
SIMS	Total (b)
Four-point probe plus thinning	Electrically active (a)
Hall measurements plus thinning (clover leaf)	Electrically active (a)
Capacitance-voltage technique	Electrically active (a)

178   Material and Device Evaluation Techniques

For a given measurement, time interval R will be a function $R(Q)$ of Q. If a thin layer, thickness Δx, is removed and the activity R is measured, $R(Q - \Delta Q)$, where ΔQ is the number of isotopes in the removed layer Δx and is the profile information. That is, using $R_0, R_1, \ldots, R_n$ to refer to the sequence of measurements,

$$Q_1 \propto R_0 - R_1$$
$$Q_2 \propto R_1 - R_2$$
$$\vdots$$
$$Q_n \propto R_{n-1} - R_n$$

Using this technique, the differential profile can be built up. Calibration of the profile can be achieved by implanting a known quantity of ions into the solid and measuring their activity. With regard to the constraint on half-life, this must be short enough to provide a measurable signal in a finite time for doping levels in the range $(10^{15} - 10^{18})\text{cm}^{-3}$ but not so short that the radioactive decay of atoms during the measurement is large compared with the initial number implanted.

A1.1.9 SECONDARY ION MASS SPECTROMETRY (SIMS)

This is a specialized technique that requires very expensive equipment (see Figure A1.8). In its simplest form, it consists of an ion beam projected onto the solid surface (Figure A1.9). The use of relatively lower energy beams of 5 to 40 keV, oxygen or cesium ions causes the surface to be eroded away slowly (this process is called sputtering). If the sputtering is controlled to occur at a given constant rate, and the concentration per unit time of the specified sputtered ions are measured by mass spectrometry analysis, then the required profile can be obtained. The sputtering time scale is converted into a depth scale by measuring the total crater depth in a given time interval. This technique also provides a relative profile that can be calibrated by using known implanted test samples.

A1.1.10 FOUR-POINT PROBE

The four-point probe technique, coupled with an anodic stripping process, can be used to determine depth profiles. Let

ρ_1 = sheet resistance before the removal of a layer
ρ_2 = sheet resistance after the removal of a layer
Δx = layer thickness removed

Fig. A1.8 Cameca 3F ion microprobe.

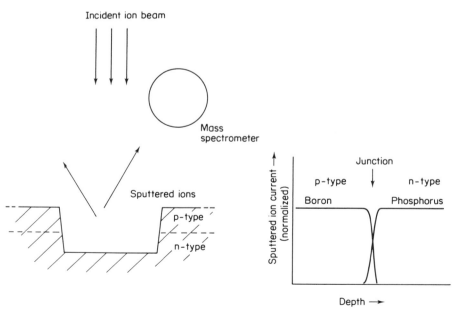

Fig. A1.9 Illustrating the principles of the secondary ion mass spectrometry.

179

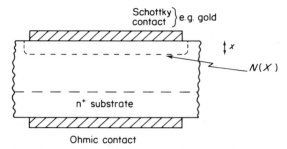

Fig. A1.10 The use of a Schottky diode for capacitance voltage profiting.

The average specific resistance of the layer removed $\bar{\rho}$ is given by

$$\bar{\rho} = \frac{\rho_1 \rho_2}{(\rho_1 - \rho_2)} \Delta x \quad \text{ohm-cm} \tag{A1.11}$$

Successive removal of thin layers Δx provides $\bar{\rho}(x)$ and, hence, the doping concentration $N(x)$.

A1.1.11 CAPACITANCE VOLTAGE PROFILING

For this technique, the layer is grown on a heavily doped substrate and a Schottky barrier is placed on the upper surface of the layer with an ohmic contact to the substrate (Figure A1.10). The basis of the technique is the change in terminal capacitance with applied voltage given by the equation

$$N(x) = - \frac{C^3}{q\epsilon_s S^2} \frac{dC^{-1}}{dV} \tag{A1.12}$$

where S is the diode area and x is given by the position of the edge of the depletion region:

$$x = \frac{\epsilon_s S}{C} \tag{A1.13}$$

Profiling information can only be obtained for distances $x > x_0$, the zero bias depletion width. The most convenient way of profiling is by using a capacitance meter or bridge to obtain the C-V relationship and to use a suitable computer program to compute the $N(x)$ versus x curve.

APPENDIX 2

Common Etches for Silicon

Composition	Comments
HF : HNO$_3$: CH$_3$COOH 3 : 5 : 3	CP4-Polishes the surface; requires 2–3 minutes.
HF : HNO$_3$: COOH 3 : 5 : 0	CPA-Polishes the surface.
HF : HNO$_3$ 1 : 3	White etch, Polishes the surface; requires 15 seconds.
Conc. HF with 0.1–0.5% conc. HNO$_3$	Delineates *pn* junctions; one drop on freshly lapped surfaces.
H$_2$O : NH$_2$(CH$_2$)$_2$NH$_2$	Does not attack silicon oxide.
NaOH or KOH 1–30% solution	Develops structural details.
HF : HNO$_3$: H$_2$O 4ml : 2ml: 4ml plus AgNO$_3$ (200mg)	Develops faults in epitaxial layers.
Ethylenediamine catechol H$_2$O (hydrazine)	Orientation and concentration dependent. Stops etching at p^{++} interface. Very slow etching of SiO$_2$.
KOH Normal propanol H$_2$O	Etches $\langle 100 \rangle$ a hundred times faster than $\langle 111 \rangle$. Stops at p^+ interface.

APPENDIX 3
Etches for Gallium Arsenide

Composition	Comments
(1–20%) Br$_2$ in CH$_3$COOH	Etching rate drops below 5%.
HF:HNO$_3$:H$_2$O 1:3:2	Good for rapid etching, can be slowed down by increasing H$_2$O.
H$_2$SO$_4$:H$_2$O$_2$H$_2$O 3:1:1	Suitable for Mesa etching.

APPENDIX 4

The Mathematics of Diffusion and Ion Implantation

A4.1 Diffusion

A simple model of diffusion is depicted in Figure A4.1 where the atomic planes are separated by a distance $d/\sqrt{3}$ where d is the spacing between the tetrahedral sites of the diamond (silicon and gallium arsenide) structure. The center of each layer contains the center of the lattice plane from which atoms may jump. If one takes the example of the silicon lattice and defines the number of atoms in each plane (1) and (2) as n_1 and n_2, respectively, then the volume concentrations N_1 and N_2 are given by (n_1/volume) and (n_2/volume); hence:

$$N_1 = \frac{n_1\sqrt{3}}{Sd}$$

$$N_2 = \frac{n_2\sqrt{3}}{Sd}$$

In this solid structure, each atom has four nearest neighboring sites it may jump into, two in the plane to the left and two in the plane to the right. Assume that one-half of the atoms may move from plane (1) to the left and one-half may move to the right. Since the single jump period is $(1/v_j)$, then the net flow of atoms from layer (1) and (2) is given by

$$\frac{\Delta n}{\Delta t} = \frac{(n_1 - n_2)/2}{(1/v_j)} \qquad (A4.1)$$

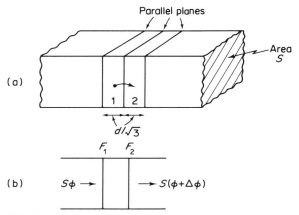

Fig. A4.1

substituting for n_1 and n_2

$$\frac{\Delta n}{\Delta t} = \frac{Sd\,V_j}{2\sqrt{3}}(N_1 - N_2) \qquad (A4.2)$$

but the concentration gradient $(\Delta N/\Delta x) = \sqrt{3}\,(N_2 - N_1)/d$ and, hence,

$$(N_1 - N_2) = -\frac{d}{\sqrt{3}}\left(\frac{\Delta N}{\Delta x}\right) \qquad (A4.3)$$

substituting into (A4.2)

$$\frac{\Delta n}{\Delta t} = \frac{-Sd^2 v_j}{6}\left(\frac{\Delta N}{\Delta x}\right)$$

Since $(1/S)\,(\Delta N/\Delta t)$ is the net flux ϕ of diffusing atoms per unit area of solid, the flux

$$\phi = -\frac{d^2 v_j}{6}\left(\frac{\Delta N}{\Delta x}\right) \qquad (A4.4)$$

The constant $(v_j d^2)/6$ is defined as the diffusion constant D and, hence,

$$\phi = -D\left(\frac{\partial N}{\partial x}\right) \qquad (A4.5)$$

This equation is called *Fick's first law* of diffusion. Consider now the diffusion coefficient $D = (d^2 v_j/6)$ in more detail. Substituting for v_j for a substitutional diffusion, D may be written as

$$D = \frac{2vd^2}{3}\exp\left[\frac{-(E_a + E_s)}{kT}\right] = D_0 \exp\left[\frac{-(E_a + E_s)}{kT}\right] \qquad (A4.6)$$

The pre-exponential term D_0 contains the vibrational frequency ν and is, hence, a function of temperature. However, as a first approximation it may be assumed that D_0 is constant over a limited temperature range (i.e., <100 K). Thus,

$$\ln(D) = [\ln(D_0)] - \frac{(E_a + E_s)}{k}\left(\frac{1}{T}\right) \qquad (A4.7)$$

A plot of the natural logarithm of D versus $(1/T)$ will yield a straight line of slope $(E_a + E_s)/k$ whose extrapolated intercept to $(1/T) = 0$ gives the value of D_0. Such a plot is shown in Figure A4.2 for aluminum phosphorus boron and antimony in silicon.

A second useful approximation is to assume that D does not depend on concentration N, but this approximation is only valid for doping densities less than 10^{20} atoms cm^{-3}. Note that for interstitial diffusions, the expression for D becomes

$$D = D_0 \exp(-E_a/kT)$$

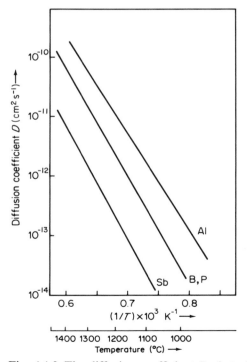

Fig. A4.2 The diffusion coefficient D plotted as a function of $1/T$ for aluminum, phosphorus, boron, and antimony in silicon.

188   The Mathematics of Diffusion and Ion Implantation

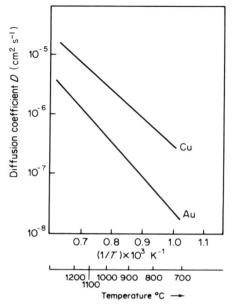

Fig. A4.3 The diffusion coefficient plotted as a function of $1/T$ for copper and gold in silicon.

Figure A4.3 shows the appropriate plot of logarithm (D) versus $1/T$ for two interstitial diffusants, gold and copper. In the comparison of curves (Figures A4.2 and A4.3), note the much lower slope of the interstitial diffusants and the much larger values of D_0.

Consider once again the simple one-dimensional model of diffusion shown in Figure A4.1b.

The flux entering face F_1 is $S\phi$ and the flux leaving at F_2 is $S(\phi + \Delta\phi)$; hence, the net flux entering the volume $(S\, \Delta x)$ is just $-(S\, \Delta\phi)$, so

$$\left(\frac{\partial N}{\partial t}\right) S\, \Delta x = -S\, \Delta\phi \tag{A4.8}$$

substituting for $\Delta\phi/\Delta x$ from Fick's first law; that is,

$$\frac{\partial \phi}{\partial x} = -\frac{\partial}{\partial x} D\left(\frac{\partial N}{\partial x}\right) \tag{A4.9}$$

$$\left(\frac{\partial N}{\partial t}\right) = \frac{\partial}{\partial x} D\left(\frac{\partial N}{\partial x}\right) \tag{A4.10}$$

which, if D is assumed to be independent of concentration, simplifies to

Fick's second law:

$$\frac{\partial N}{\partial t} = D \frac{\partial^2 N}{\partial x^2} \qquad (A4.11)$$

This differential equation is very important, as its solution, subject to specific boundary conditions, enables controlled diffusion profiles to be obtained and *pn* junctions to be fabricated. In most practical conditions two boundary conditions allow relatively simple solutions to Fick's law to be used.[1] These are discussed in Chapter 3.

A4.2 Ion Implantation

The basic theoretical equations that are used to predict the range distribution of ions implanted into solids were developed by the Danish group at the University of Aarhus (Lindhard Schaff and Schiott and are known as the LSS theory). In this theory, the following assumptions are made:

a. The solid is assumed to be amorphous with no open channels to allow the implanted ions to become trapped and, hence, to have anomalous deep ranges (i.e., channeling).
b. The incident ion beam is mono-energetic and does not contain a mixture of isotopes.

In real single crystals, the phenomenon of "channeling" occurs. In this phenomenon the ions becomes trapped in the open planes and axes and can have ranges of many times the normal. Although it is hard to completely inhibit this efect, it can be minimized by misorientating the incident beam with respect to the open axial and planar channels.

The LSS theory predicts that the range distribution $N(x)$ will be approximately a Gaussian (Figure 3.16) with a projected range $\overline{R}_p$ where

$$N(x) = \frac{\phi}{\sqrt{2\pi}\sigma_p} \exp\left| -\frac{(x - \overline{R}_p)^2}{2\sigma_p^2} \right| \qquad (A4.12)$$

Where ϕ is the total number of ions implanted (fluence cm^{-3}) and σ_p the standard deviation in projected range.

Note that at $x = \overline{R}_p$ the peak concentration $N(x) = N_p$ becomes

$$N_p = \phi/(\sqrt{2\pi})\sigma_p \simeq 0.4\phi/\sigma_p$$

---

[1]Those who wish to pursue this topic in more detail, refer to Appendix 6, reference 1.

The total number of ions implanted is

$$\phi = \int_{x=0}^{\infty} N(x)dx$$

that is, the area under the implanted profile curve.

A comprehensive outline of the theory of Lindhard and co-workers is given by Gibbons et al., who list the numerical calculations for a range of ions in silicon and gallium arsenide. Some of these data are plotted in Figures A4.4a and b.

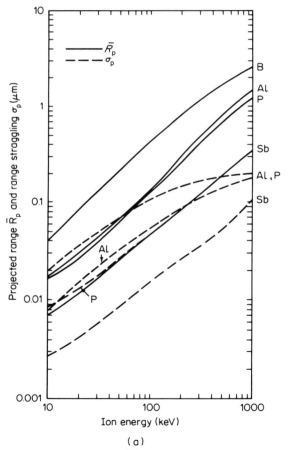

Fig. A4.4 Range data $\bar{R}_p$ and σ_p for some p- and n-type dopants in (a) silicon and (b) gallium arsenide.

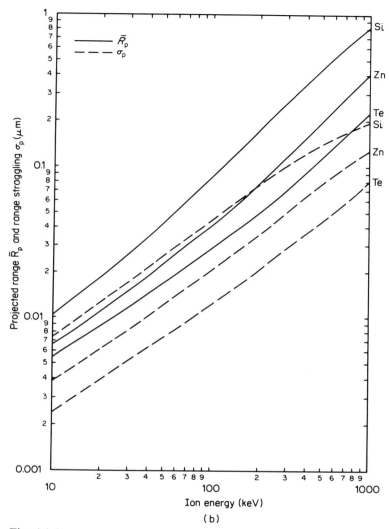

Fig. A4.4

REFERENCES

J. Lindhard, M. Scharff, and H. E. Schiott, Mat. Fys. Medd Vid. Selek **33**, 14 (1963).

J. F. Gibbons, W. S. Johnson, and S. W. Myeroie, *Projected Range Statistics: Semiconductors and Related Material*, Dowden, Hutchinson and Ross, Stroudsburg, PA, 1975.

APPENDIX 5

Thermal Oxidation of Silicon

Consider a unit-area cross section of semiconductor plus oxide in which the oxidant concentration N (molecules per cubic centimeter) varies linearly from N_s at the surface to N_i at the interface, as is shown in Figure A5.1. If the flux of oxidant molecules passing through the oxide satisfies Fick's first law of diffusion (see Appendix 4),

$$F_0 = -D_0 \frac{dN}{dx} = \frac{D_0 (N_s - N_i)}{x_0} \tag{A5.1}$$

The flux F_i represents the rate at which the oxidant molecules are consumed at the interface in the production of SiO_2. It is reasonable to assume that $F_i \propto N_i$:

$$F_i = k_i N_i \tag{A5.2}$$

In the steady state, the two fluxes must be equal ($F_0 = F_i = F_f$, say). Equating (A5.1) and (A5.2),

$$N_i = \frac{N_s}{1 + k_i x_0/D_0}$$

and

$$F_f = \frac{N_s}{1/k_i + x_0/D_0}$$

If N_m oxidant molecules are required to produce unit volume of SiO_2,

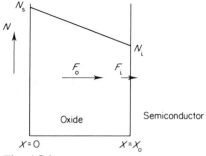

Fig. A5.1

then the rate at which the SiO$_2$ film thickness increases is given by

$$\frac{dx_0}{dt} = \frac{F_f}{N_m} = \frac{N_s/N_m}{1/k_i + x_0/D_0}$$

Integrating from 0 to t with $x_0|_{t=0} = x_i$ gives

$$x_0 - x_i + \frac{k_i}{2D_0}(x_0^2 - x_i^2) = \frac{N_s k_i}{N_m} \quad \text{(A5.3)}$$

Eq. A5.3 is a quadratic in x_0 and may be solved to give

$$x_0 = \frac{A}{2}\left[\sqrt{\left(1 + \frac{t + t_0}{A^2/4B}\right)} - 1\right] \quad \text{(A5.4)}$$

where $A = 2D_0/k_i$, $B = 2D_0 N_s/N_m$, $\tau_0 = (x_i^2 + Ax_i)/B$.

APPENDIX 6
References and Bibliography

REFERENCE	COMMENT
1. H. M. Olsen, Chapter 2 (*Device Technology*) in *Variable Impedance Devices*, M. J. Howes and D. V. Morgan (eds.), Wiley, New York, 1978.	An excellent chapter dealing with *pn* diodes and Schottky diodes. Good on GaAs.
2. S. K. Ghandi, *Theory and Practice of Microelectronics*, Wiley, New York, 1968.	A comprehensive text covering technology and device design. Excellent coverage, good reference book.
3. R. A. Colclaser, *Microelectronics*, Wiley, New York, 1980.	Very good text. Wide coverage and up to date.
4. P. E. Gise and R. Blanchard, *Semiconductor and Integrated Circuits Fabrication Techniques*, Prentice-Hall, Englewood Cliffs, N.J., 1979.	A good laboratory style book—very useful to the beginner.
5. P. Daniels. Chapter 5 in *Large Scale Integration*, M.J. Howes and D. V. Morgan (eds.). Wiley, New York, 1981.	A concise and up-to-date survey of technology.
6. S. M. Sze, *The Physics of Semiconductor Devices*, 2nd Edition, Wiley, New York, 1981.	One of the most comprehensive semiconductor books and an excellent reference book.
7. E. H. Rhoderick. *Metal Semiconductor Contacts*. Oxford University Press, New York, 1958.	A comprehensive and up-to-date text on this topic. To be recommended to the advanced student.
8. G. Carter and W. A. Grant, *Ion Implantation in Semiconductors*., E. J. Arnold, Leeds, U.K., 1976.	A good introductory book.
9. A. Grove, *Physics and Technology of Semiconductor Devices*, Wiley, New York, 1967.	Excellent treatment of technology although now dated.

10. T. P. Kaberservice, *Applied Microelectronics*, West, St. Paul, Min., 1978. — Good section on electronic factors in LSI.
11. E. S. Yang, *Fundamentals of Semiconductor Devices*, McGraw-Hill, New York, 1978. — Excellent up-to-date text of device theory and technology.
12. C. Mead and L. Conway, *Introduction to VLSI Systems*, Addison-Wesley, Reading, Mass., 1980. — This has become the standard text on chip level design.
13. J. Watson, *Semiconductor Circuit Design*, 4th Edition, Adam-Hilger, Bristol, U.K., 1983. — Broad based introductory text that includes the transition to circuit design.

BIBLIOGRAPHY

Mitra, S. K., *An Introduction to Digital and Analog Integrated Circuits and Applications.*, Harper & Row, New York, 1980.

Neudeck, G. W. and Pierret, R. F., *Modular Series on Solid State Devices*, Vols. 1 to 4, Addison-Wesley, Reading, Mass., 1983.

Pierce, J. F., *Semiconductor Junction Devices*, Charles E. Merrill, Columbus, Ohio, 1967.

Richman, P., *MOS Field-Effect Transistors and Integrated Circuits*, Wiley, New York, 1973.

Runyan, W. R., *Semiconductor Measurements and Instrumentation*, McGraw-Hill, New York, 1975.

Streetman, B. G., *Solid State Electronic Devices*, 2nd Edition, Prentice-Hall, Englewood Cliffs, N.J., 1980.

van der Ziel, A., *Solid State Physical Electronics*, 2nd Edition, Prentice-Hall, Englewood Cliffs, N.J., 1968.

Veronis, A., *Integrated Circuit Fabrication Technology*, Reston, Va., 1979.

APPENDIX 7

Supplemental Exercises

1. Write down a general expression for the conductivity of slab of a uniform semiconductor containing both electrons and holes.
   How does this expression simplify for a sample of:
   (a) intrinsic material, and
   (b) extrinsic n-type material?

2. Figure A7.1 shows a plot of the logarithm of conductivity versus $1/T$ for a semiconductor—what can you deduce from this?

3. A bar of silicon contains 2×10^{24} boron atoms m^{-3} and 3.5×10^{24} antimony atoms m^{-3}. Using Figure 3.7 determine the mobilities of the electrons and holes (μ_n and μ_p) and the resistivity of the bar.
   Why does the resistivity of the bar differ from one containing just 1.5×10^{24} antimony atoms m^{-3}? The data in Fig. 3.7 may be used for your calculations.

4. Using a suitable figure define what is meant by the sheet resistivity.

5. Figure A7.2 shows the structure of a commercial GaAs transferred electron device and also a plot of the device current ratio $J_{B=0}/J_B$ versus the square of the magnetic field B^2 for a device biased below threshold (directions as shown in the figure). Show with the appropriate equations what you may deduce from this.

6. Define the term "relaxation time (τ)" for a charge carrier moving in a semiconductor and discuss factors which control τ in real semiconductors.

7. Given a slice of semiconductor material in the form shown in Figure A7.3a and the experimental information shown in Figure A7.3b, calculate the conductivity, doping density, nature of semiconductor (p- or n-type) and mobility. (Derive all equations used.)

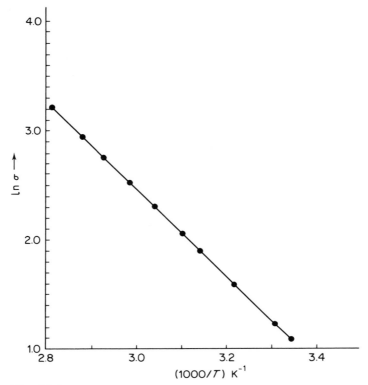

Fig. A7.1

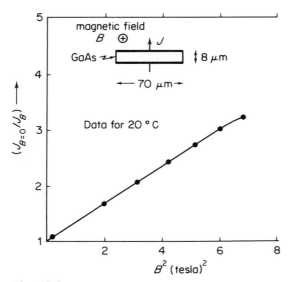

Fig. A7.2

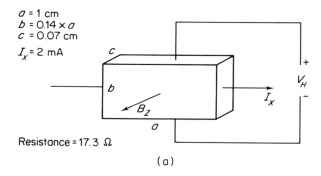

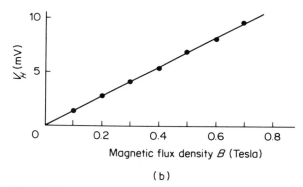

Fig. A7.3

8. In an experiment on a semiconducting specimen of resistivity 0.05 Ω m, the Hall coefficient was found to be 45×10^{-3} m^3 C^{-1}. What are the free charge carriers, what is their concentration, and what is their mobility?

9. The electrical conductivity of a piece of intrinsic material doubles when its temperature is raised from 300 K to 310 K; discuss fully how you could use this information to calculate the band gap E_g of the material (discuss any equations and approximations used in your calculations). (*Note*: $kT = 0.025$ eV at 300 K.)

10. The equilibrium current through a slice of *n*-type germanium is increased by 0.1% when it is illuminated by a light that forms hole-electron pairs at the rate of 2×10^{22} m^{-3} s^{-1}. Calculate the lifetime (resistivity = 1.74 Ω m and mobility $\mu = 1.74 \times 10^{-1}$ m^2 V^{-1} s^{-1}).

11. Discuss the basic assumptions leading to the conclusion that one-dimensional diffusion processes in a semiconductor may be described by

$$\frac{\partial N}{\partial t} = D \frac{\partial^2 N}{\partial x^2}$$

where N is the chemical concentration of the diffusing atoms.

12. Describe the physical mechanism by which diffusion of dopant atoms may occur in silicon (restrict your discussion to the so-called substitutional dopants).
    Diffusion is governed by Fick's second law:
    $$\frac{\partial N}{\partial t} = D \frac{\partial^2 N}{\partial x^2}$$
    Discuss briefly the solutions to this equation for (a) constant source diffusion and (b) limited source diffusion. Illustrate schematically how the diffusion profiles so obtained will vary with temperature and time. (*Hint*: The diffusion coefficient $D = D_o \exp[-(E_a + E_s)/kT]$.)

13. A *p*-type silicon sample has a uniform doping density of $N_A = 10^{23}$ atoms m^{-3}. Determine the diode junction depth when the sample is subjected to the following process: Predeposition with excess phosphorus at 950° C for 30 minutes (i.e., note that the solid solubility concentration of phosphorus at 950° C is 8×10^{20} atoms m^{-3}). Plot the resulting diffusion profile.
    Figure 4.2 is a plot of D versus $(1000/T)$ for a range of atoms in silicon. The table is a listing of Z versus erfc (Z)

Z	0	0.094	1.808	2.713	3.617	4.520
erfc (Z)	1	0.2	0.01	1.23×10^{-4}	3.1×10^{-7}	1.6×10^{-10}

14. A donor impurity is diffused into a semiconductor that has a nondiffusing background acceptor impurity (concentration 10^{21} m^{-3}), from a constant source (concentration 10^{23} m^{-3}). Given that the diffusion time is 10 hr and the diffusion temperature 1100° C, plot the resultant doping profile through the semiconductor. You may assume that the activation energy of the diffusion process is 3 eV and the apparent value of the diffusion coefficient (of the diffusing atoms) at $T = \infty$ is 10^{-2} m^2 s^{-1}.

15. What are the advantages of ion implantation over diffusion for device fabrication?

16. Describe why the occurrence of a natural oxide (SiO_2), with its excellent electrical and physical properties, was so important to the development of silicon planar technology and, hence, to silicon integrated circuits.

17. Discuss in detail the process of thermal oxidation of silicon, indicating the basic mechanism involved in the process.

18. By assuming that the surface concentration of oxygen is governed by its solid solubility limit N_0 and using a linear approximation for the concentration profile, show that the relationship between oxide thickness x and time t is given by the relationship
    $$x = \frac{D}{k}\left[\left(1 + \frac{2N_0 k^2 t}{Dn}\right)^{1/2} - 1\right]$$
    where D is effective diffusion constant for the oxygen in the oxide, n the number of molecules of oxygen per unit volume of the oxide, and k the reaction rate constant.

19. What are the advantages and disadvantages in using an in-contact mask to form patterns in photoresist?

20. Explain why positive resists produce images rather larger, and negative resists produce images smaller than the mask dimensions.
21. Given that a surface alignment mark has several microns of epitaxial silicon grown over it, explain how the next masking level can be aligned to it.
22. Design a mask set for two junction diodes connected back to back.
23. Describe the advantages and limitations of masking to higher levels of integration in the design of electronic systems.
24. Discuss the relative merits of wet and dry thermal oxidation.
25. Compare the times taken to grow a 1 μm silicon oxide film at 1100° C using
    (a) a dry oxide,
    (b) a steam-grown oxide.
26. An oxide film is grown for 1 hr at 950° C in (a) dry oxygen, (b) steam. Calculate the resulting thicknesses and comment.
27. What are the limits to the optical photolithography process when fabricating submicron device structures? Describe in detail how these limitations may be overcome.
28. Describe how the process of photolithography is used in microtechnology and using the example of a positive resist show with clear diagrams how the process may be used to fabricate mesa diodes on an oxide-coated silicon wafer.
29. Starting with a uniformly doped silicon epitaxial layer ($N_D \simeq 10^{23}$ m^{-3}) 10 μm thick grown on an n^+ substrate ($N_D > 10^{24}$ m^{-3}) and using the data in Figure A4.4, determine a scheme to produce a p^+n junction where the metallurgical junction is 0.2 μm below the surface. Discuss the reasons for your choice of p-type dopant and the procedure needed to activate the implanted ions.
30. Outline the detailed technological steps and sketch the masking schemes needed to fabricate a MESFET transistor suitable for operation up to 10 GHz starting from:
    (a) an n-type epitaxial layer on a semi-insulating substrate,
    (b) a uniform semi-insulating substrate. Start by specifying your initial material requirement. (*Hint*: Study Figure 7.21.)
31. Compare the relative merits of the two diode structures illustrated in Figures 7.19*b* and *c*.

APPENDIX 8
Properties of Silicon, Gallium Arsenide, and Silicon Dioxide

Property	Si	GaAs	SiO$_2$
Atoms or molecules per centimeter^3	5.0×10^{22}	2.21×10^{22}	2.3×10^{22}
Atomic or molecular weight	28.08	144.63	60.08
Density, g/cm^3	2.33	5.32	2.27
Breakdown field, V/cm	$\sim 3 \times 10^5$	$\sim 3.5 \times 10^5$	$\sim 6 \times 10^6$
Crystal structure	Diamond	Zinc blende	Amorphous
Dielectric constant	11.8	10.9	3.9
Electron affinity, χ,v	4.01	4.07	0.9
Energy gap, eV	1.12	1.43	~ 8
Intrinsic carrier concentration, cm^{-3}	1.5×10^{10}	10^7	
Lattice constant, Å	5.431	5.654	
Effective mass:			
Electrons	$m_e = 0.33$ m,	$m_e^* = 0.26$ m	0.068 m
Holes	$m_h = 0.56$ m,	$m_k^* = 0.38$ m	0.56 m
Intrinsic mobility:			
Electron, cm^2/V-sec	1350	8600	
Hole, cm^2/V-sec	480	250	
Temperature coefficient of expansion	2.5×10^{-6}	5.8×10^{-6}	5×10^{-7}
Thermal conductivity, W/cm C	1.45	0.46	0.01

INDEX

Acceptors, 39
Activation energy, 51
Alloying, 105
Anisotropic etches, 33, 181
Annular saw, 31
Atomic diffusion, 50–56, 120, 124, 185–191
AuGe, 105
Automatic compensation, 56

Ballistic motion, 164
Band diagram, 38
Batch processing, 2, 3, 8
Beam lead, 113
Bipolar process, 150
Bipolar transistors, 124, 142, 150
Bonding, 107
Buried layers, 126

CAD, 155
Capacitance voltage profiling, 180
Capacitors, 135
CCDs, 131, 165
Charge coupled devices, 131, 165
Chemical etching, 32, 181, 183
Chemical vapor deposition, 28
CMOS, 146
Compensation, 44
Complemenary error function, 53
Conductivity, 39
Constant source diffusion, 53
Contacts, 10
Cost per gate $vs.$ time, 161
Crystal growth, 15–29
Crystal orientation, 29
Crystal purification, 15
Crystal structure, 18
CVD, 27, 28, 81
CVD reactor, 27, 72
C-V measurements, 80
Czochralski growth, 18

DAC, 153
Deep levels, 48

Depletion FET, 129
Deposited oxides, 71, 80, 193
Design rules for thermal SiO_2, 73
Design sequence, 153
Diamond saw, 31, 110
Diamond scriber, 110
Dicing, 107, 110
Diffusion, 10, 50, 121, 124, 185
Diffusion-rate limited growth of SiO_2, 78
Diffusion theory, 185–191
Digitizing, 95
Diode capacitors, 135
DMOS, 147
Donors, 40
Doping profiles, 177
Doping of semiconductors, 37, 56
Drive in diffusion, 54, 65
'Dry' thermal oxide, 73

E-beam evaporation, 99–100
E-beam lithography, 163
ECL, 143, 155, 158
Electroless plating, 103
Electrolytic plating, 103
Electron beam lithography, 163
Enhanced growth of SiO_2, 78
Enhancement FET, 129
Epitaxial growth, 22
Etches for gallium arsenide, 183
Etches for silicon, 32, 181
Evaporation filaments, 100
Extrinsic semiconductors, 39

Failure rate, 162
Fast interface states, 79
Ficks Law, 52, 186, 188
Field oxide, 149
Film quality assessment, 80
First reduction, 93
Fixed interface charge, 79
Flip chip, 113
Float zone process, 20
Four point probe, 171, 172, 178

Freeze-out temperature, 42
Future technology developments, 162

Gallium arsenide, 14, 136
Gallium arsenide etches, 183
Gate-drain capacitance, 145
Gate-source capacitance, 145
Geometric magnetoresistance, 175
Gigabit logic, 165
GMR, 176

Hall coefficient, 174
Hall effect, 174
Hall mobility, 174
HCL-effect on film quality, 78
HCL-effect on SiO_2 growth, 78
Hermetic sealing, 11
High pressure oxidation, 80
Horizontal BJTs, 126

ICs gallium arsenide, 136
I^2L, 144, 158
Impatts, 104, 106, 113
Impurity redistribution, 79
Initial artwork, 93
Instantaneous source diffusion, 54
Insulating films, 69
Integrated circuits, 1
Integration, 2–6, 153
Interstitial atoms, 40
Interstitial diffusion, 48
Intrinsic carrier density, 38
Intrinsic semiconductors, 38
Ion beam milling, 35
Ionic charge, 79
Ion implantation, 56, 121, 130, 136, 189–191
Ion implanter, 62

JFET, 126
Jump frequency, 52
Junction FET, 126

Kerf, 108

Lap and stain, 170
Large volume requirements, 6
Laser scribing, 110
Lift off, 123
Light emitting diodes, 165
Liquid phase epitaxy, 23
Lorentz force, 174
Low-resistance contacts, 10

LPCVD, 72
LSI, 160

Magnetoresistance mobility, 175
Masking for diffusion, 70
Masking for ion-implantation, 70
Masking levels, 10
Mask making, 85
Masks, 85
Material evaluation, 169
Mean time between collisions, 44
Meandered resistors, 133
Melbourne puller, 21
Mesa diodes, 118, 119
MESFETs, 127, 139, 163, 165, 177
Metal-gate NMOS, 145
Metallization, 97–106
Mobile ionic charge, 79
Mobility, 39, 44, 174
Mobility-Hall, 174
Mobility-magnetoresistance, 175
MOCVD, 27
Molecular beam epitaxy, 25
Monolithic diode, 120
Monolithic ICs, 165
Monolithic resistors, 133
Morphology, 169
MOS, 69, 80, 129, 141
MOS capacitor, 136
MOS transistor, 142
Mounting, 107
Mylar, 94

n-Channel, 145
Negative resist, 88
New semiconductors, 164
Nitride films, 81
NMOS, 145

Ohmic contacts, 98, 104, 139
Overexposure, 89
Oxidation, 10, 71
Oxidation systems, 73, 74
Oxidation theory, 193
Oxide films, 71

p-Channel, 146
Parabolic rate constant, 78
Passivation, 69, 107
Pattern formation, 9, 85
Photolithography, 10, 85
Photoresist, 88

Photoresist spinner, 91
Photoresist thickness, 90
Planar diode, 120
Planar technology, 159
Plasma deposition, 29
Plasma etching, 33
Plating, 103
PMOS, 146
PN junction, 118
Polysilicon, 90, 144
Positive resist, 88
Practical oxidation systems, 73, 74
Predeposition, 63
Projected range, 57
Prototype integrated circuit, 8

Q factor, 136
Quasi-intrinsic semiconductors, 44

Radiation damage, 56
Radioactive isotype profiling, 177
Range data, 190–191
Range straggling, 57
Reaction-rate limited growth of SiO_2, 78
Recombination, 41
Reduction camera, 95
Reliability, 159
Resistivity, 169, 171
Resistors, 133

Sawing of Wafer, 10
Scaling, 163
Schottky contacts, 104
Schottky defect, 52, 56
Schottky diodes, 123, 180
Self-alignment, 145
SEM, 170
Shadow etch technique, 132
Sheet resistance, 171
Sidewall definition, 90
Silane, 71
Silicon etches, 181
Silicon gate, 145
Silicon-gate NMOS, 145, 148
Silicon nitride, 82

SIMS, 177, 178
SiO_2, 70, 73, 193–194
Sodium contamination, 80
Solid solubility, 53
Sputtering, 35, 101
Step and repeat, 95
Straggling data, 190–191
Submicron developments, 163
Substitutional diffusion, 50
Surface passivating layers, 69

Technology families, 142, 158
Thermal oxidation, 193–194
Thermal oxides, 72
Thermasonic bonding, 111
Thermocompression bonding, 111
Thin film deposition, 98
Transferred electron device, 110, 139
TTL, 158
Type conversion, 37, 49

Ultrasonic bonding, 11, 111
Underexposure, 89
Unwanted impurities, 47

Vacancy, 51
Vacuum evaporation, 98
Van der Pauw sample, 175
Vapor phase epitaxy, 23
Varactors, 58
Vertical BJTs, 124
'V' groove etching, 33, 34
Vibrational frequency, 51
VLSI, 163
VMOS, 131, 147

Wafer thinning, 106
Wet thermal oxide, 73
Wire bonding, 10

X-ray lithography, 86

Yield, 7, 162

Zone refining, 15